Mike McRae started his science career more than twenty years ago in a hospital medical laboratory. With a passion for education he took to teaching science in London's east, followed by a year entertaining children across Australia with Questacon's Science Circus. Mike has been writing science for over a decade, working with the CSIRO, the ABC and the Australian Museum to produce educational materials, to inform and to entertain. Today he writes for the popular online news service ScienceAlert, and contributes regularly to a variety of science news and education publications. Mike is the author of *Tribal Science: Brains, Beliefs & Bad Ideas*.

WHAT MAKES
A DISEASE
A DISEASE?

# UNWELL

## MIKE McRAE

UQP

First published 2018 by University of Queensland Press
PO Box 6042, St Lucia, Queensland 4067 Australia

uqp.com.au
uqp@uqp.uq.edu.au

Copyright © Michael McRae 2018
The moral rights of the author have been asserted.

This book is copyright. Except for private study, research, criticism or reviews, as permitted under the *Copyright Act*, no part of this book may be reproduced, stored in a retrieval system, or transmitted in any form or by any means without prior written permission. Enquiries should be made to the publisher.

Cover design by Christa Moffitt
Author photo by Alex Massey
Typeset in 11.5/15 pt Bembo Std by Post Pre-press Group, Brisbane
Printed in Australia by McPherson's Printing Group, Melbourne

The University of Queensland Press is assisted by the Australian Government through the Australia Council, its arts funding and advisory body.

A catalogue record for this book is available from the National Library of Australia

ISBN 978 0 7022 6031 5 (pbk)
ISBN 978 0 7022 6181 7 (pdf)
ISBN 978 0 7022 6182 4 (epub)
ISBN 978 0 7022 6183 1 (kindle)

University of Queensland Press uses papers that are natural, renewable and recyclable products made from wood grown in sustainable forests. The logging and manufacturing processes conform to the environmental regulations of the country of origin.

# CONTENTS

**PART I BROKEN   1**
Nostalgic   3
Sick   12
Hysterical   19
Sinister   26
Fragile   31
Diseased   41

**PART II INFECTED   51**
Invaded   53
Contaminated   61
Fermented   66
Inflamed   77
Inhabited   88
Clean   94

**PART III INSANE   101**
Addicted   103
Bad   115
Psycho   127
Guilty   132
Perverted   138
Ashamed   151

**PART IV ABNORMAL   157**
Dirty   159
Cut   170
Fat   185
Weak   199

**PART V DEAD 207**
Rotten   209
Expired   220
Arrested   230
Suicidal   239
Human   251

**PART VI WELL   259**
Extraordinary   261
Good   271
Fertile   283
Fit   293
Valuable   301

*Acknowledgements   309*
*Notes   313*

*For Liz*

I
BROKEN

# 1
# BROKEN

# NOSTALGIC

Sorry if I don't remember your face. I have this condition, you see, where a tiny part of my brain doesn't work as you would expect. It goes by the name prosopagnosia, a word that comes from the Greek words for 'face' and 'ignorance'.

I share this minor disability with a famous neurologist, the late Oliver Sacks, who once admitted: 'Several times I have started apologising to large, clumsy, bearded people and realise that it's a mirror.' I don't think I'm quite as bad as that, but I have mumbled to myself as a child approaches, 'That boy is wearing the same clothes as my son' – only to realise a moment later that it is in fact my son.

That's not all that's wrong with me. I also lose hope sometimes, a disposition that colours my world in charcoal, misery and ennui. Psychologists call this problem chronic depression. They suggest I exercise more, take pills and treat myself to a movie once in a while. Some months are better than others. I also get bad bronchitis after suffering a cold. And I get mouth ulcers when I use some types of toothpaste.

My wife is a little broken too. She finds it hard to concentrate sometimes, and so leaves cupboard doors open and doesn't put the pickles back in the fridge. They call it attention deficit disorder (the kind without hyperactivity), and she takes little white tablets of methamphetamine to help her focus. Her forgetfulness can bug me sometimes, but it's all part of her charm.

Then there's my son. He often paces and claps his hands, he won't look you in the eye much, and he speaks in loops and idioms. You might think that makes him a little odd. But he can recite all the US presidents in order, he's brilliant with numbers and computer games, he's always laughing and he calls a spade a spade. If you like, you can call his broken bits autism.

Most of my friends are broken. Some are left-handed – but nobody really cares about that anymore, do they? Some take medicine to calm their anxiety or ease their depression. I have a father who, throughout my childhood, had a compulsion to drink too much alcohol. I have a brother who was born with a tiny extra digit jutting from the thumb on one hand. It was cut off when he was still a baby, just in case.

Most of us are broken in our own way. But if my family and friends lived at other moments in history, or in other parts of the world, it's possible we'd be considered more than simply 'broken'. We'd be melancholic, forgetful, crazy, idiotic, perverted, weak, lazy or selfish. We'd be bad. We'd be sinful. Stupid, if not plain sick. Even if physically we felt fine and dandy, morally we'd be corrupt. Our treatments might be exile, incarceration or beatings. Or worse.

When should our differences be considered weaknesses? When do our weaknesses become pathologies? And how significant do our odd quirks need to be before we're labelled as diseased and disabled?

I have twin nieces with pancreases that don't make enough insulin. I had a friend who died of bowel cancer. I had chickenpox when I was eleven and glandular fever when I was nineteen. We would find it strange for anybody to question if these were true diseases. Yet hardly a month goes by without somebody in the media asking – rhetorically, and often leadingly – where we should draw the line on particular afflictions.

'Is ADHD a real disease?' asked *The New York Times* in 2011.[1] Back in the 1990s they wanted to know: 'Gulf War Syndrome – Is it a real disease?'[2] 'Is obesity a disease or a choice?' said a voiceover promoting a documentary on SBS.[3] The *Daily Mail* was also curious: 'Is addiction an illness or a weakness?'[4] *Forbes* wondered about 'Restless legs syndrome – a pharma creation or a real health problem?'[5] In light of the sexual assault scandals shaking Hollywood in 2017, *Time* turned to the experts for an answer to the question: 'Is sex addiction real?'[6]

For decades the press has loved to challenge us with these kinds of questions. And who can blame them, really? We, their audience, are desperate to know who deserves our pity and who our criticism. Is it Chronic Fatigue Syndrome or plain laziness? Is it oppositional defiance disorder or are they just little buggers who don't like to follow orders? Is it gluten intolerance or are they precious millennials who think it's fashionable to have a snowflake diet? Mental illness or religious extremism? Personality disorder or asshole?

Among the lists of diseases and disorders we've catalogued over the centuries, it's not hard to find mistakes. There are numerous sicknesses long abandoned like pathological ghost towns, leaving unanswered the question of whether they were illnesses that met a cure or medical mirages. Some contagious diseases, such as smallpox, legitimately vanish as we drive their agents towards extinction. Others, such as leprosy or scurvy, are legacies of history we're unlikely to encounter thanks to improved hygiene and nutrition. Then there are conditions that never were, such as neurasthenia – a disease some rather smugly called 'Americanitis'.

Back when the United States was still healing from the ravages of civil war, its citizens staggered bloody into a modern age of golden opportunity, celebrating the idea that anybody who worked hard enough could succeed at life. In the late 1800s America's population exploded to over 70 million as immigrants flocked to its shores in search of health, wealth and happiness. Major cities swelled as poor rural families sought fortune in revolutionised industries. America had become electrified, mechanised, mass-produced and fast-paced. It also became anxious, irritated, fatigued and impotent. This new America witnessed the rise of a new disease, one popularised by a psychologist named William James.

Americanitis manifested as anything from headaches, palpitations and high blood pressure to weakness, indigestion, neuralgia, loss of focus and depression. So wide and varied were the symptoms that it was one of the most common diagnoses given to patients between the years 1880 and 1920. Whatever ailed you, this disease was probably the cause.

The blame was placed squarely on the speed of life: the hustle and bustle, the crash-bang chaos of the modern world. 'Hurry is stamped in the wrinkles of the national face,' inspirational author Orison Swett Marden wrote in his book *Cheerfulness as a Life Power*; he whimsically advised against worry in order to prevent suffering from this most debilitating disease. The neurologist Silas Weir Mitchell referred to the disease as a 'disorder of capitalist modernity'. An 1898 article in the *Journal of the American Medical Association* accused the loud calamity of urban life: 'who shall say how far the high-strung, nervous, active temperament of the American people is due to the noise with which they choose to surround their daily lives?'[7] There were also suggestions that the malady was caused by the new electrical lights, which enabled longer workdays; still others believed distance from nature was making it hard for city-dwellers to revitalise their depleted energy reserves.

The more clinical and sanitised term for Americanitis was 'neurasthenia' – an old-fashioned word that once broadly described weak nerves. Whatever you called it, it was considered a bona fide nervous condition. It was not until 1980 that it was finally struck from the grand tome of psychological disorders, *The Diagnostic and Statistical Manual of Mental Disorders* (generally known as the *DSM*).

These days, it's unlikely a doctor would take your pulse, listen to your woes and tell you they've never seen such a chronic case of neurasthenia. Either we don't suffer from nerves in the same way or Americanitis was not a *real* disease. Yet for nearly half a century it was something just about everybody seemed to suffer. Who can we blame for making such an outrageous categorical error?

The origins of the condition can be traced back to a Connecticut-born neurologist, George Miller Beard, who in the late 1860s put flesh on the bones of what was basically anxiety. He wrote extensively on the condition, believing it to be epidemic among hardworking, urban, white Protestants of the northern states. The characteristics of those diagnosed with the disease spoke volumes about how society defined 'healthy' as well as 'afflicted'. Catholics, according to Beard, were relatively immune from the religious anxiety that plagued other sects, since they left all their spiritual concerns to a hierarchy of theologians. Similarly, Beard believed life among the agrarian white southerners was simple enough to not stress their nerves, while those of African heritage were of such 'immature mind' that they were hardly at any great risk of stress and anxiety.

Far from being stigmatised as a frailty, neurasthenia was rather fashionable. Villains in mid-nineteenth-century stories were often depicted as insane with syphilis. The poor and tragic were weak with consumption. Yet the heroes of the tales – those industrious, salt-of-the-earth Protestants facing up to the challenges of the brave new world – were often worn down, drained and neurasthenic.

This is not to say that the symptoms of neurasthenia – anxiety, listlessness, melancholy, loss of appetite, impotence and a generally lacklustre attitude towards life – were being fabricated by the alleged sufferer. Americanitis wasn't an empty box, but an attempt to explain a disparate variety of symptoms expressed by people who were desperate for a name to give their devil. Nor was Beard the first to attempt to clinically diagnose depression or anxiety. He was, however, the first to attempt to use the burgeoning field of neurology to explain why people felt out of sorts.

According to his explanation, babies slipped out into the world with a set quantity of nervous energy. Once this supply ran dry, the body became depressed, irritated and mentally exhausted. A link between electricity and the functioning of nerves had been established earlier in the nineteenth century; in an increasingly industrialised world, it's easy to see how a metaphor of nervous power that could be drained and replenished might appeal.

Treatments were as varied as the symptoms, and fanned the fervour behind this imagined condition. The disease was a boon for apothecaries and snake-oil salesmen alike. Newfangled electrical stimulation devices claimed to calm and reinvigorate. Elixirs and tonics with names such as Dr Miles' Nervine, Weller's Nerve Vitalizers and Neurosine became popular pharmaceuticals, possibly thanks to ingredients like cannabis and alcohol. Even good old

Coca-Cola – in its original, non-carbonated form – claimed to treat this ubiquitous disorder of the nerves.

Men were typically told to get out and exercise more. For those who could afford it, the prescription was to venture west, with forays into the untamed wilds just the ticket to cure the impotent, anxious, disillusioned modern American male. Perhaps the most famous of neurasthenic sufferers was the future US president Theodore Roosevelt, who, on being accused of being too effeminate as a result of his nervous condition, took to the western wilderness to toughen up.

For women, on the other hand, the cure was total withdrawal from the world, or at least a break from constant cooking and cleaning. Some cases were considered to be so severe that physicians commanded complete bed rest in a dim room, without so much as a conversation or even a book to alleviate the soothing boredom. Yet in an age of rising suffrage, feminists such as the novelist Charlotte Perkins Gilman argued for domestic duties to be shared in cases of neurasthenia: women were wasting their nervous energy on frivolous pastimes, and not contributing to society.

In reality, neurasthenia was more to do with what was expected of the modern Protestant American than with poor health. It spoke to the cultural standards of masculinity and femininity that citizens struggled to meet, expectations that risked being abandoned in the name of progress. Americanitis was the noble sacrifice of leisure and happiness for the moral good of the newly established, free, hardworking middle class. Humans were built for nature, not cities, and disease was the price we paid.

It's not that Americans weren't feeling sad, impotent or dispassionate. There's every possibility that being surrounded by strangers in a time-focused society contributed to higher stress levels, much as it does today. But the need to have a simple biological model to blame – one that meant your suffering was legitimate and you weren't a bad person – demanded the creation of a catchy name. Neurasthenia meant you could be weak and tired without it being your fault. The world had changed, and your God-given body just wasn't suited to this unnatural landscape. Change, apparently, isn't always as good as a holiday.

In fact, back in the seventeenth century, sadness over a change in surroundings was responsible for another serious, even deadly, condition that also later vanished from the medical books.

The Swiss physician Johannes Hofer believed that the homesickness – or *mal du pays*, as he called it – of the Swiss mercenaries who were touring the lowlands of France should be thought of as a serious illness; he said it was a 'neurological disease of essentially demonic cause'.[8] He combined the Greek terms *nostos* for 'homecoming' and *algos* for 'pain' to form the new word *nostalgia*, which has of course come to describe a much less diabolical (if still sometimes heartbreaking) condition.

Illness from loneliness was hardly a revolutionary idea. Hofer was the first to paint nostalgia as a true disease, and his idea would influence nearly two centuries of medical diagnosis. While aboard the *Endeavour*, the botanist Joseph Banks wrote in his diary that the sailors 'were now pretty far gone with the longing for home which the physicians have gone so far as to esteem a disease under the name of nostalgia'.[9]

Nostalgia was more than just a name for the pain over missing Mum's lamb roast. It was a brutal condition that risked debilitating physical symptoms. In his dissertation *Considérations sur la nostalgie*, the nineteenth-century French physician François-Théophile Collin argued that nostalgia was a legitimate disease 'since it can terminate in death caused by a series of latent inflammations, and, in as little as a month's time, attacks various organs and weakens all the functions, above all those of the nervous system'.[10]

Jean Baptiste Félix Descuret, in his 1841 book, *La médecine des passions*, records case studies of sufferers of nostalgia. One tells of a recluse who was forced to leave his much-loved condemned home. 'Two days later,' he writes, 'the commissar was required to force open the door of the stubborn lodger. He found the poor man dead; he had suffocated from the despair of having to leave the abode he cherished too much.'

Nostalgia was one of a variety of conditions that French physicians called *maladies de la mémoire*, representing a period in which it had become fashionable to study the pathology of memory. The disease continued to be seen as a serious illness until the mid-nineteenth century – about the time neurasthenia was gaining notoriety in the United States – before fading into a gentle pining for home.

Maybe better technology allowed nostalgics to connect with home through faster travel. Maybe they suffered less because they could now swap photographs and eventually words through 'instantaneous' communication devices, such as telegraphs and telephones. Is it possible that nobody dies from nostalgia anymore because nobody is more than an email away these days? Perhaps. Or maybe nostalgia, like neurasthenia, was never a real disease in the first place.

There's no doubt that people through history have felt the painful sting of abject loneliness in a foreign land, or that they've at times suffered from a loss of appetite, insomnia and general misery. Especially when vast distances made them feel more isolated. For that matter, Americans – indeed, citizens from all over the industrialised world – have certainly experienced all manner of aches, pains, anxieties, weaknesses and distress in periods of rapid social upheaval following a time when the world was much simpler.

But these two conditions, neurasthenia and nostalgia, were more than a line drawn around symptoms in a changing world. They described a specific relationship between biology and society – a failure of the body and the mind to meet expectations, resulting in suffering. As time wore on, expectations shifted and the lines were erased and redrawn elsewhere.

Today you can't claim 'nostalgia' on Medicare, and your boss probably won't accept 'neurasthenia' as the reason you couldn't come in last Monday. And despite the fact our world hasn't slowed down, we don't now lock women in dark rooms when they feel like they can't manage the housework, or send presidents out on bear hunts if they seem too effeminate. With hindsight, we might be tempted to see these 'conditions' as the fumbling attempts of physicians who just didn't know any better. But how many modern conditions might one day meet the same fate? Obesity a disease? Ludicrous! Addiction? No way. ADD, ODD, CFS, ASD, PTSD? What were those fools thinking back then?

Science can provide us with invaluable insights into the variety of genes, biochemistries and neurological wirings in our global population. There's no doubt that, as we learn more, many things we think of as conditions in today's world will develop in complexity, and we'll redraw their borders, combining some and abandoning others.

But as we go about this never-ending task, we'll still be relying on fundamental assumptions about what it means for human biology to be broken, as if diseases lie undiscovered like new lands ready for mapping. We know the cartographers of illness can make mistakes, but what if the terrain is itself an illusion, and by drawing lines we're inventing as much as we're discovering?

# SICK

I remember when I first heard about 'Typhoid Mary'. I was in high school, and was thrown by the question of whether or not she had a disease. More than just an illustration of infection, she's the perfect mix of crime story, medical case study and morality tale. It was Mary, many decades ago, who first inspired me to question the nature of disease.

The back-of-the-book version of her story goes like this: Mary Mallon was an Irish migrant to the United States in the late nineteenth century, and in her gut she carried a microbe called *Salmonella typhi*. In most people this nasty little bacterium would result in typhoid, characterised by a fever, rash and diarrhoea. About a fifth of people who don't receive treatment will die. Mary, though, didn't show a single symptom. But she did spread the bacteria on practically everything she laid her hands on, infecting dozens of people as she moved from job to job as a cook. She was locked up because of her tainted touch, following her discovery by a private investigator. After thirty years of being incarcerated, held in isolation, Mary Mallon died of pneumonia at the ripe old age of sixty-nine.

Biology fails. As complex organic systems, we are subject to the grinding wear of entropy, and so we feel weak, sad, sore and uncomfortable in countless ways; ultimately, nature doesn't really give a damn. And we use words like *disease, sickness, illness, disability, syndrome, disorder* and *injury* to describe these states. Sometimes we use these terms interchangeably, but each also implies a unique way we think about our wellbeing.

So how is a *disease* different to a *disorder* or a *syndrome*? What's the relationship between the concepts of *sickness* and *disability*? *Illness* and *injury*? Are there obvious distinctions? Does it even matter?

There's no shortage of historical opinions. 'Disease is an abnormal state of the body which primarily and independently produces a disturbance in the normal functions of the body,' said the medieval Persian polymath Avicenna.[11] At the other end of the timeline, the late-nineteenth-century American pathologist William Thomas Councilman claimed: 'Disease may be defined as "A change produced in living things in consequence of which they are no longer in harmony with their environment".'[12]

To really know what a disease is, we first have to know what it means to be healthy and functional. In 1946 the World Health Organization (WHO) offered a working description for health as 'a state of complete physical, mental and social well-being, not merely the absence of disease or infirmity', meaning you could argue you were unhealthy even if nobody could pinpoint a specific medical problem.[13] Health, according to the WHO, is a matter of feeling comfortable and relatively happy, which can dissipate even if we otherwise seem functional.

Disease, we might therefore conclude, is a state of function that makes us less than comfortable. If you flick through any one of the several bazillion textbooks on disease published every year, you'll mostly find variations on this theme. Disease is some sort of physical or mental state that contrasts with equally vague concepts like *harmony*, *wellbeing* and *normality*.

That's the textbook stuff out of the way. As tempting as it is to hold aloft the dictionary or defer to some other authority, let's close those heavy dictators of language and consider instead how we – ordinary people – actually use the words.

For most of us, it's easier to define *disease* by examples and their causes. Measles is a disease caused by a virus. So are chickenpox, herpes and mumps. Acquired immune deficiency syndrome (AIDS) is a disease caused by the human immunodeficiency virus (HIV). Tetanus is a disease caused by a bacterial infection, and so are syphilis, typhoid and tuberculosis. Thrush is caused by a fungus, and malaria is caused by single-celled parasites invading red blood cells. All of these diseases are 'caught' as infectious agents move into or around our bodies and reproduce.

Cancer is the product of mutated genes that have caused cells to rapidly divide. Type 1 diabetes is the result of a dysfunctional pancreas. Alzheimer's is caused by changes to our central nervous system. Each of these examples of disease has one or more functions we point to as out of order, resulting directly or indirectly in the loss of comfort.

*Illness*, on the other hand, lacks the need for diagnosis. You can call your boss and say you're ill without telling them your pharynx is inflamed due to a streptococcal infection. Symptoms are all you need – a sore throat, a headache, an upset stomach.

Consider now how *disorders* imply physical characteristics we are born with. And how you can't typically heal from conditions we describe as *disabilities*. There is no 'former' state of normality to return to for these things. They aren't merely injuries. As uncomfortable as they are, broken bones, bruises and cuts aren't diseases or disorders. If pressed, we might describe diseases as things we 'catch', disorders as problems we are born with, and injuries as physical tears and breaks resulting from trauma. We can heal from an injury or a disease in ways we can't heal from a disorder.

A *disability* is a disorder that compromises the senses or functions we expect our bodies to have. Some could be born with this physiological difference, while others might lose the feature by way of an incident later in life. Either way, negotiating any interpersonal or physical environments is made more challenging thanks to society's expectations about what it means to be functionally able.

Having so many words to describe biological malfunctions is how we impose order on what is essentially a broad and remarkably complicated system. Most of the time, words like these work fairly well. In fact, our modern perspective of disease and disability arose because medicine successfully returned countless people to a satisfactory state of health and wellbeing.

Unfortunately, biology isn't always so straightforward. Agreeing on clear boundaries for terms like *wellbeing* and *functional* is virtually impossible, thanks to the fact that we can't always agree on what makes an ability valuable in the first place. Even settling on whether a bunch of symptoms makes up one disease or two can be a nightmare.

That one name has described multiple diseases – or, conversely, that one disease has had multiple names – has made cataloguing diseases a challenging task. Just ask the British epidemiologist William Farr, who wrote back in 1839 in the first Annual Report of the Registrar-General of England and Wales:

> The advantages of a uniform nomenclature, however imperfect, are so obvious, that it is surprising no attention has been paid to its enforcement in Bills of Mortality. Each disease has, in many instances, been denoted by three or four terms, and each term has been applied to as many different diseases: vague, inconvenient names have been employed, or complications have been registered instead of primary diseases.

Farr wasn't alone in believing that the creation of a universal, consistent system for identifying and naming diseases was a matter of the highest priority, on a par with efforts to standardise other fields of research. 'The nomenclature is of as much importance in this department of enquiry as weights and measures in the physical sciences,' he wrote, 'and should be settled without delay.'

Naming, defining and cataloguing diseases has been a serious business ever since. Florence 'Lady of the Lamp' Nightingale – the world's most famous nurse alongside Margaret 'Hot Lips' Houlihan – might be remembered for her eye for cleanliness, but it was her way with numbers that really changed the world. Nightingale added empirical rigour to hospital accounting by presenting an analysis of patient care to the 1860 International Statistical Congress in London, paving the way for a new field of disease statistics. Thirty years later, a French statistician, Jacques

Bertillon, developed a mortality classification system, cataloguing life-threatening maladies that targeted specific parts of the body.

The end result is the modern *International Statistical Classification of Diseases and Related Health Problems* (or *ICD*, for less of a mouthful). Currently in its eleventh revision, the *ICD* is the go-to compendium for international health organisations, and the closest thing we have to a complete list of human diseases and disorders. We categorise how we feel and function by slotting symptoms into its relatively tidy boxes. Different countries adapt the classification system to suit the peculiarities of their own medical cultures, but each version records diseases in the same way, using a coding system that associates related diseases and advises on diagnoses and treatment procedures. The cerebral cousin to the *ICD* – the American Psychiatric Association's *Diagnostic and Statistical Manual of Mental Disorders* (or *DSM*) – has been in use since the middle of the twentieth century. The two catalogues have worked hard to align their lists, attempting to use the same codes and provide consensus on their respective diseases of the mind. Together, these mighty tomes represent the absolute authority on what ails you.

It's tempting to try to count the codes to come up with a precise number of diseases known to humanity. I don't recommend doing this. The nature of the task quickly becomes clear as codes expand into sub-codes, making it impossible to know whether a single description constitutes one disease or several. Where do you start? How far down do you stop?

To make matters worse, different national versions of the *ICD* also present the diagnosis of a specific disease code in subtly different ways, with unique modifications. This includes the addition of what are known as culture-bound syndromes – symptoms considered a disease in one place but not another. Take the condition amok, for example, which has the code F68.8 in the clinical modification version of the *ICD-10*. Few countries would have a use for the category, though in English we use the term as an idiom to describe unrestrained activity. Among Malay and Indonesian cultures, however, *amok* is a violent and frenzied outburst that often follows a period of brooding. While it was once considered the result of a demonic possession, today it's officially a psychiatric condition in a handful of South-East Asian countries.

The dynamic histories of the *ICD* and *DSM* reflect just how impossible the defining and cataloguing of diseases can be. The World Health Organization's most recent disease compendium, *ICD-11*, was released in June 2018. The current fifth edition of the *DSM* was approved by the American Psychological Association's board of trustees in 2013, and doubtlessly there will be a sixth in due course.

Diseases have come and gone from both catalogues as diagnostic technology has progressed and discoveries have been made, as different cultures have successfully argued for their own diagnoses, and as fashions have demanded a change in language or a destigmatising of a function. To sum up, we just don't have a simple list of all the diseases known to humanity. One wonders whether Farr would've been impressed or distressed by today's state of disease curation.

The authorities of illness – those cartographers of disease – have a tough job finding the right balance as they map and remap the pathological landscape. I don't envy them. But if they struggle, how are we non-physicians to make sense of our health?

Putting it simply, the line between health and sickness isn't the result of objective, physical laws. There are no simple rules we can apply to answer those tabloid questions of whether alcoholism or sexual appetite is a disability or a weakness in willpower. Disease is largely subjective, often conforming to our individual cultural experiences.

We like the idea that the work of a medical scientist is similar to that of a palaeontologist, seeking the hidden bones of disease amid the surrounding matrix of biology. But we're wrong. To sum up the state of our understanding, we can't go past the words of Harvard historian Charles Rosenberg: 'Disease is at once a biological event, a generation-specific repertoire of verbal constructs reflecting medicine's intellectual and institutional history, an aspect of and potential legitimation for public policy, a potentially defining element of social role, a sanction for cultural norms, and a structuring element in doctor/patient interactions.'[14] Maybe it's not all that snappy, but Rosenberg captures the difficulty of defining disease perfectly.

Could it be that my friends and family aren't that broken after all? While there's no doubt there are pathogens and mutated genes that cause suffering, we can't dismiss the fact that, for most of us, a disease is whatever our doctor diagnoses. Beyond that, why should we even care?

Most of the time we trust the authorities to make decisions to keep us safe by removing the Typhoid Marys from our midst. This is a most wonderful thing. Unless, of course, you happen to be poor old Typhoid Mary.

# HYSTERICAL

Between my legs is a pair of testicles. They're nothing amazing, but I'm rather attached to them, and embrace them (metaphorically) as a sign of my relative masculinity. I'm privileged to have them, because throughout history those two olive-sized organs would have entitled me to be a passionate, argumentative, opinionated person who could express ideas like those in this book, and not have people question my mental health for doing so.

Now, if I had a fist-sized lump of flesh known as a uterus, I might not be so lucky. Especially if I were to somehow time-travel back a century or so. My testicles might occasionally jiggle a touch, and even shift up and down a little, but they don't stray far from their anchorage and cause me to go a bit ... well, nuts.

As far back as ancient Greece, anatomists believed it was possible that many of our internal organs could fall out of place and wander about the body aimlessly, like grapes in a fruit salad. You can hardly blame them for thinking the squishy lumps deep inside our torso weren't fixed in place – cutting up fresh cadavers wasn't a weekly event, and when some ancient doctor did get a good peek inside the occasional gladiatorial loser, the organs weren't always where he expected them to be. Even once dead people became more freely available for study, many physicians refused to let go of the theory that some of our organs caused problems by relocating from their usual spot.

The mythical wandering uterus – now better known as an outdated disorder called hysteria – has been considered a problem in the West for at least 2000 years. Esteemed medics such as Aretaeus of Cappadocia and Hippocrates of Kos claimed that an inflamed uterus could slip its moorings and enter the chest cavity, for example, where it risked suffocating the poor owner. After all, why else would a member of the fairer sex suddenly change from a quiet and demure creature into an emotional, sexually charged being? Aretaeus even referred to the womb as an 'animal within an animal', given its wild and mobile nature, and prescribed coaxing it back into place with a whiff of certain fragrances.

As late as the nineteenth century, a handful of doctors stubbornly clung to the belief that the uterus (as well as the kidneys, spleen, liver, bowel and other giblets) could sag and drop in what they called visceroptosis. A French physician named Frantz Glénard figured droopy organs were behind that other notorious disease of yesteryear, neurasthenia. Thanks to his fame, visceroptosis was also referred to as Glénard's Disease.

Robert Battey, a young surgeon from the US state of Georgia, was such a fan of cutting out functional 'droopy' gonads that his name became synonymous with ovariotomy operations. Short of removing wobbly organs – or stitching them in place, as one pioneering German doctor liked to do – kinder physicians suggested tight bandages might be a safer option. Others thought constraining garments were to blame, claiming corsets were pushing organs around and causing pain, elevated blood pressure, loss of appetite and constipation.

Aside from surgery and binding, a misbehaving uterus could be jarred back into place through much more pleasant forms of treatment. British surgeon Sir Arthur Hurst prescribed exercise and Swedish gymnastics, before finally deciding later in life that the whole notion of dropped organs was bunkum.

Then there's the good, old-fashioned orgasm. The influential Roman physician and anatomist Galen thought women became emotional when they had too much semen gumming up their wombs. The excess needed to be expelled, and that could be done by inducing a muscle-clenching climax, which would force the womb to return to its place, balance out the fluids and perhaps even – with luck – end in a pregnancy. Way more fun than forcing

out a sneeze or two, like the boring old Byzantine Paul of Aeg
recommended.

By the Middle Ages, the idea that a uterus would dramatically rise and fall through the abdomen was abandoned. For a few centuries at least. The uterus was still known to be the source of all manner of female ailments, though. And stimulation remained a cure for women behaving badly.

Take the word of the seventeenth-century 'Dutch Hippocrates', Pieter van Foreest, who devoted a whole chapter to women's diseases and disorders in his treatise.[15]

> When these symptoms indicate, we think it necessary to ask a midwife to assist, so that she can massage the genitalia with one finger inside, using oil of lilies, musk root, crocus, or similar. And in this way the afflicted woman can be aroused to the paroxysm. This kind of stimulation with the finger is recommended by Galen and Avicenna, among others, most especially for widows, those who live chaste lives, and female religious, as Gradus proposes; it is less often recommended for very young women, public women, or married women, for whom it is a better remedy to engage in intercourse with their spouses.

Note his use of the term *paroxysm*. Bodice-rippers might use it as a euphemism for joy, but make no mistake: it's not a sexy word.

By the nineteenth century, it was firmly believed that women did not – indeed, could not – orgasm as men did. The heavy-breathing, eye-rolling, lip-biting, muscle-clenching fit was not an erotic, thrilling rush, according to a prudish establishment. The female paroxysm was more akin to epilepsy than pleasure, and therefore it was perfectly acceptable for a male physician to offer a helping hand. The belief that sexual gratification of any kind could only be produced by an erect penis reinforced the idea that stimulating the clitoris digitally wasn't a sexual act.

In case you were thinking this was a corrupt attempt by virile young medically trained (and trusted) men to slip their fingers up a skirt or two, the task was universally reviled. Massaging a clitoris was time-consuming and tiring work, done only for the purposes of treatment.

Necessity being the mother of invention, hand-cramped doctors employed a tool invented by an English physician named Joseph Mortimer Granville. Known as 'Granville's hammer', the electric percussion device was designed to massage sore muscles and improve memory. No, really. The fact that physicians found it useful for inducing paroxysms was not a purpose that pleased the sceptical Granville. 'I have never yet percussed a female patient,' he once wrote.[16] Perhaps he considered the idea vulgar, but more likely he just didn't think hysteria was a real disease. 'I have avoided, and shall continue to avoid the treatment of women by percussion, simply because I do not wish to be hoodwinked, and help to mislead others, by the vagaries of the hysterical state.'

Nonetheless, other doctors didn't think it was so vague, and they went on to design similar machines powered by electricity or even steam, to buzz, shake, massage and rock their female patients, in order to save their precious digits from unnecessary effort. These appliances soon made their way into the home market, and were advertised in popular women's magazines such as *Woman's Home Companion* as 'personal massagers'. Not that anybody was fooled; the ads promised 'all the pleasures of youth' that would 'throb inside you'.

Within half a century, another invention would strip the subtlety from this personal massager and give it a less wholesome reputation. With the advent of film came a new art form – motion picture pornography. There was no disguising the vibrator when its true purpose became a popular stag-party entertainment, and women's magazines quickly stopped advertising them. With the rise of moral conservatism in many parts of the West, laws were passed in the mid-twentieth century to prohibit the sale of anything that might be intended for sexual stimulation.

The idea of hysteria as a disorder was slow to fade, but the stereotype of the hormonal, emotional woman has barely dimmed. In some part, the blame for women regularly behaving in unacceptable ways has only migrated – from the uterus to the endocrine system.

Similar to hysteria was a disease called chlorosis, also commonly referred to as 'green sickness' or 'the disease of virgins'. For half a millennium, young women who were a little too irritable were diagnosed with this illness. The seventeenth-century English physician Thomas Sydenham, often called the 'English Hippocrates', depicted what he referred to as 'virgin's pale colour' as a kind of hysteria consisting of the following symptoms:[17]

1   a bad colour of the face, and whole body
2   a swelling of the face, eyelids and ankles
3   heaviness of the whole body
4   a tension and lassitude of the legs and feet
5   difficult respiration
6   palpitation of the heart
7   pain in the head
8   feverish pulse
9   drowsiness
10  an unnatural longing for such things as are noxious and unfit for food
11  a suppression of menstrual discharge.

His prescription was for the patient to drink mineral water from an iron mine. By the early twentieth century the condition had become entwined with iron-deficiency anaemia, and ultimately the two diseases merged.

But that doesn't mean the vanishing of chlorosis was just a name-swap. If we look at it closely, it's clear chlorosis was more than low iron levels (if it was ever even that to begin with). Once again, the origins of the disease can be found with those ancient know-it-alls, the Greeks. This isn't all that coincidental – for much of the Middle Ages, European physicians were of the opinion that there were no diseases unknown to those sages of medicine. When presented with a list of symptoms, it was just a matter of digging through old texts and finding a match.

So when the sixteenth-century German physician Johann Lange was asked by a friend to diagnose his daughter, Anna, Dr Lange turned to Hippocrates, who quite literally wrote the book on women's diseases. It was in the recently translated *On Diseases of Virgins* that Lange found what he was after. In it Hippocrates warned of the risks young women could face if they didn't menstruate on cue. An excess of blood due to 'food and growth of the body' couldn't escape; it pooled up, pushed on the heart and diaphragm, and caused all sorts of problems. To sort this out, women needed to get pregnant.

Anna purportedly had a pale face, 'as if bloodless'. Lange went on to note that 'her heart trembles severely at any bodily movement, and the arteries of her temples pulsate with feeling; she has laboured breathing when dancing or climbing stairs; she avoids food, especially meat; her legs – in particular near the ankles – swell at night'. Clearly, this had to be the 'disease of virgins'.

The physician suggested a treatment of herbs that were thought to encourage menstruation. He also figured bloodletting might do the trick – or, better still, getting hitched. In his letter to his friend, Lange reported that he'd gladly attend the wedding. Anna must have been thrilled.

Bawdy seventeenth-century songs cry of an illness 'which many maids do call the sickness green', and report that they would do anything 'for a dil doul'.[18] (Which is to say, most likely a penis, hinting at the possible origins of the word *dildo*.) As with hysteria, the cure for chlorosis was to get that uterus working properly, which, as we now know, requires an orgasm or three.

Hysteria and chlorosis were fundamentally reflections of a false premise: that intercourse was a male sport. Even though most Europeans during the Middle Ages accepted that women enjoyed sex as much as (or even more than) men, within a few hundred years the dominant belief emerged that the female body was built solely to receive sperm and germinate an embryo. Sexual gratification played no obvious role in baby-making. Women could obviously enjoy being penetrated, however for three-quarters of the female population climax requires a touch more stimulation than intercourse.[19]

In a world where sex is still often considered taboo, female orgasms appear to be an anomaly, and therefore abnormal. Since the authoritative books are overwhelmingly written by people who don't own female genitalia – and who don't think it important to ask questions like 'So, what's this paroxysm business like?' – a few anatomical assumptions are bound to creep in.

Hysteria was a disease only when women did not act as society expected them to. Errant female behaviour became a condition, one that could be managed by a medical authority, always male. And that authority's ideas about what a woman's body should do, not how it operated in reality, turned functions now considered quite normal into diseases.

# SINISTER

Sitting in a cradle on a set of bookshelves in my study is a decades-old replica of a medieval long sword. Its claim to fame is that its blade was forged by Peter Lyon, the swordsmith who, in later years, would build an armoury for director Peter Jackson's Tolkien trilogies. There are a bunch of nicks and scratches along its edges, left by collisions with other weapons. Yeah, I was one of those ultra-geeky historical re-enactors.

More recently, I gave sabre fencing a go. I can't say I'm a good sword fighter, as the dings in my helm, shield, head and little finger can attest to. If I were left-handed I might have had some small advantage, as I discovered when touring Scotland in my twenties.

Ferniehirst Castle sits outside of the Scottish town of Jedburgh. The buildings on the estate have experienced the decay and reconstruction typical of most historical European castles, repurposed over the centuries to suit the needs of its occupants, so the original structures are now virtually hidden in plain sight. It doesn't even look classically medieval. But the vestiges of at least one set of features have remained in place since their addition in the early sixteenth century. Ferniehirst Castle's spiral staircases wind the wrong way.

To climb the steps in a typical medieval tower, you spiral anticlockwise, putting your right hand close to the outer wall as you ascend and your left hand over the void. This is no accident of architecture – imagine you're launching an assault on the castle, blade in your right hand clanging awkwardly against the stonework, the castle's guard defiantly holding the staircase from above. In such narrow confines, winding up for a bone-crunching blow is made all the harder with a wall in the way, while the defender has an easy time swinging their right arm down onto your head.

Ferniehirst Castle must have seemed like a welcoming invitation for right-handed raiders. But its flipped staircases retained an advantage for one particularly sinister laird of the estate and his fellow mercenary clansmen. Born way back in 1471, Dand was the firstborn son of the Kerr clan chieftain. He was said to have been brave, ruthless and quick-tempered. And left-handed. Traditionally, being left-handed has all the hallmarks of a disability. To Dand Kerr and his band of mercenaries, however, it offered an advantage. One I can attest to.

Most right-handed combatants, who outnumber southpaws something like nine to one, would have relatively limited experience fighting left-handed opponents. Left-handed fighters could hire out their sword arm for twice as much as their more common peers. So the Kerr chieftain trained his men to fight left-handed, and built his spiralling staircases to suit their unique style. In 'The Raid of the Kerrs', the Scottish poet James Hogg immortalised their talent:[20]

> But the Kerrs were aye the deadliest foes
> That ever to Englishmen were known
> For they were all bred left handed men
> And defence against them there was none.

If there's any truth at all to this story, there's little reason to believe that Dand happened upon a bunch of congenitally left-handed mercenaries. So, just as learning to be right-handed has never been a walk in the park for lefties, it couldn't have been easy for these mostly right-handers to adapt. I've tried swapping hands for the slight advantage left-handedness might have offered in armed combat, and still have the scars to prove how hard it is.

To fit into a right-dominated world, many who have a dominant left side will attempt to learn to use their weaker hand. In many Eastern nations, only around 1 to 4 per cent of the population demonstrate left-handedness – a significant difference to the 10 per cent found elsewhere in the world – indicating that there is either something significantly different in the physiology of humans across the world, or there's been a whole lot of hand-swapping happening in certain cultures.[21] An older study, conducted on a small group of adolescents living in the Amazon, found there were zero lefties, leading to speculation that it was universally trained out at a young age so that all could use weapons and other communal tools.[22]

Many Eastern countries continue a practice that older folk in the West might remember: the use of encouragement and punishment to force children to use the 'correct' hand. And it's not just a few. A 2007 study on Taiwanese students, for example, found that nearly two-thirds of students were still being coerced into using their right hand.[23]

Looking around my desk right now, it's clear I live in a right-handed world. The scissors are angled to require the tilt that only a right hand can easily produce. My computer mouse favours those with a dominant right hand. Even my writing pad has sentences flowing left to right, so I can avoid dragging my hand through the ink as I scribble down notes. Approximately 10 per cent of the population is forced to hunt for knives, scissors, can openers, fencing foils, guitars, computer gadgets and even guns that suit their 'evil' handedness.

There's evidence that those who do manage to convert from mostly using their left hand to their right hand have persistent signs of left-handedness in the premotor and parietal areas of their cerebral cortex.[24] Whatever the ultimate cause (or causes) of this lateral dominance in the brains of humans, it's safe to say it isn't the result of lifestyle choice or cultural fad. Once a leftie, always a leftie.

Laboratory studies on chimpanzees suggest that roughly a third of the apes show a preference for their left side.[25] Conversely, parrots (such as the cockatoo) seem to be mostly left-handed.[26] In any case, a preference for using one limb over another is not an exclusively human trait.

It's still not entirely clear what causes a species or individual to have a dominant hemisphere of their body. Since signs of handedness in an embryo is a pretty good indicator of its handedness later in life, it seems the bias is embedded long before birth. Genetics might play some indirect role, as there is a slightly greater chance that left-handed parents will have a left-handed child. There are also hints that the influence of androgen hormones on certain genes while we're in the womb might have an effect,[27] especially since men are slightly more likely to be left-handed than women.

Historically speaking, having a dominant left hand has been more than just an inconvenient quirk of physiology – it's been a sign of inferiority and immorality. The word *left* comes from the Anglo-saxon word *lyft*, meaning 'weak'. In Latin the word is *sinister*, a term that has also evolved to mean evil. Compare that to *dexter* for right, from which we get *dexterous* – a word that means nimble and skilled. 'Left' in Sanskrit is *waama*, once again a word that implies evil. In French the word is *gauche*, which also carries a sense of being wrong, incorrect and ugly. In fact, in just about every language the word for 'left' carries implications of badness, wrongness, clumsiness, weakness or moral imperfection.

It gets worse. In the latter half of the twentieth century, some scientists even correlated left-handedness with a bunch of moral and health problems. In 1977 an American forensic psychologist named Theodore Blau argued that left-handed folk were slightly more likely to suffer behavioural and academic challenges.[28] (To be fair, he also conceded that lefties could be more imaginative than righties.) Shortly after, a Harvard University neurologist, Norman Geschwind, suggested that left-handers suffered more migraines and learning disorders than right-handers.[29] The American neuropsychologist Stanley Coren famously argued that lefties lived shorter lives.[30] So not only do southpaws have a hard time finding scissors they can use, they're also apparently low-IQ deviants who'll probably drop dead sooner than the rest of us.

On closer inspection, left-handers probably have little to worry about. Coren based his findings on two investigations. The first used a ready source of hand preference – encyclopaedia entries on baseball players – although follow-up studies on the same data failed to reach the same conclusion. His second data source was a survey sent to families in southern California. Here, he found the average age of death for right-handers was seventy-five, while it was just sixty-six for left-handers.

In 1993 the University of Leicester psychologist Marian Annett pointed out a simple reason for this that doesn't mean left-handed folk are disadvantaged.[31] Prior to World War II, the percentage of left-handers in the US population was about 3 per cent, as most 'converted' to right-handedness. This rose to about 10 per cent in the second half of the century as the social pressure to conform eased off. The mean age of the population of left-handed folk therefore fell as the decades rolled on and fewer children were forced to feign right-handedness. Taking this decreasing shift in average age into account, Coren's survey statistics look far less damning.

Not only does Annett's critique give left-handers one less thing to worry about, it also shows how important it can be to consider the role culture plays in how we define disease and disorder. Being left-handed isn't a moral or even a physiological disorder. It's just a variation that we've simply not liked all that much in the past.

The hand with which we're most coordinated — which we use to throw balls, write letters and cut food — is a physical part of our bodies from birth. Its status as an unwanted abnormality in need of correction has largely been the result of cultural values that find it odd. Swordsmen like Dand Kerr might have taken advantage of this, but for schoolchildren in Taiwan, being left-handed has been a disability. It's the morality of the authority that has made it a 'bad' thing.

# FRAGILE

Chlorosis, hysteria, neurasthenia, nostalgia and wrong-handedness are relatively outdated conditions. They're terms that can join other abandoned medical language, including dropsy, spastic, retarded, lockjaw and grippe. Medicine rolls on, leaving behind that which becomes tainted or insufficient. What of new diseases, then? If conditions can die, how are they born?

Let me tell you about Joyce McRae.

Joyce was born in the heat of a Brisbane summer in 1923, the youngest of three children of a Scottish immigrant named David and his wife, Elsie. On 10 December 1949 Joyce married a salesman by the name of Anthony 'Tony' Farmer, saying 'I do' in the Ann Street Presbyterian Church. The couple moved into a house they built at Mount Gravatt. They had two children, who would go on to produce a generation of five grandchildren. Of these, I was the first.

In her later years, a number of brain cells in Joyce's substantia nigra began to die off, and her basal ganglia degenerated, producing lower amounts of a neurotransmitter called dopamine. This gave her shaky hands — which she cheekily called the 'ta-tas', as if she was forever waving goodbye — as well as increased signs of depression, hypokinesia and rigidity. In her seventies she was diagnosed with a disease named after British physician James Parkinson.

But this wasn't the disease that took Joyce's life. In 2008 she had a fall. Her pelvis broke. And, in hospital, the complications of internal bleeding were too much for her frail, 85-year-old body.

You won't find my grandmother's name in any medical journal or on Wikipedia. She's not famous. In fact, the cause of her death is so common, so ordinary, that it's often simply accepted as a fact of ageing. Statistics gathered by the Australian Institute of Health and Welfare suggest that around 90 per cent of all hip fractures occur in people over the age of sixty, with women outnumbering men by more than two to one.[32] Most fractures were the result of a fall, likely following a trip or a stumble. Another report indicated that about a quarter of those who sustain a pelvic fracture don't survive another twelve months.[33]

It's likely my grandmother died because her bones were so brittle: a fall that might merely have bruised a younger body for her resulted in a mortal fracture. Inside more youthful bones, cells called osteoclasts are constantly dismantling bone tissue, while other cells – osteoblasts – go about rebuilding it. This constant bone renovation remains fairly balanced until our forties or fifties, when it slowly starts to dismantle more and rebuild less. The integrity of the bone gradually deteriorates, and the risk of fractures climbs.

Brittle bones have always gone hand-in-hand with ageing. That old people get frail isn't news. Yet this kind of fragility has only been considered a disease within the past generation or so. Had my grandmother known her fate as a little girl, she would have seen it as the inevitable outcome of her twilight years. These days, the story of her passing is more complicated: Parkinson's disease made it hard for her to recover from an imbalance, and a brittle bone disease called osteoporosis made her pelvis more prone to breaking.

Half a century ago, osteoporosis wasn't a disease. In 1968 an article in the journal *The Lancet*, in response to an ongoing debate regarding fractures in the elderly, argued that there was 'little justification for continuing to regard senile or postmenopausal osteoporosis as a disease, distinct from the loss of bone with age'.[34] With the support of statistics and references to numerous studies, the authors claimed that bone loss was a normal part of ageing, and that fractures occurring as a result of bone brittleness should be considered typical.

Fast-forward several decades and the WHO came to a very different conclusion, in a report exploring the factors involved in correlating bone mass density with risks of fracture.[35] While the report didn't comment directly on the debate, this was a defining moment for osteoporosis as a disease. Once an authority had drawn a line in the sand, physicians could measure the density of a person's bones, consider their age, gender and environment, and determine whether or not they had a disease that could be treated.

Osteoporosis is just one example of a condition of biology that has graduated from being considered a mere variation in form and function to a fully fledged disease. What exactly changed?

Science, for one thing. Over the decades, we discovered more about the underlying biochemistry of bone growth and the impact of factors such as diet and hormones. This meant we could not only paint a clearer picture for individuals, but make decisions that just might reduce the rate of bone loss, and therefore reduce the likelihood of someone developing a fracture. Osteoporosis took on qualities that we readily associate with diseases: we learned about the variation in bone density, re-evaluated our expectations and identified ways to impede the condition's development.

No doubt other, more subtle cultural factors have contributed to our drawing of a line between disease and natural decline. Milk production and distribution companies benefit from people consuming more calcium, for example, and have actively promoted the consumption of dairy products as a way to combat osteoporosis. While that's not to imply that osteoporosis is in any way a *cow*spiracy, marketing dollars like this can contribute significantly to the general public's recognition of a disease.

Today, a range of organisations are devoted to promoting awareness and treatment of osteoporosis. As a disease, it has a status it lacked when it was considered a normal part of ageing. Special attention is devoted to identifying risks and providing resources for researching treatments.

*Diagnosis* is the cornerstone of modern medicine. Specialised tests measure the comparative concentrations of hormones, nutrients and electrolytes, the shapes and sizes of cells and tissues, and the colour and arrangement of organs. In my earlier days working in a pathology laboratory, staining slides, inserting tubes into machines and holding my breath as I sifted through faecal and urine specimens in search of parasites or salt crystals, I played a part in the process of identifying why a person felt unwell. By doing this, physicians could come up with a *prognosis* (a prediction about what will happen to the patient under certain conditions) and a *treatment* (an attempt to deviate from the fate biology has provided).

The reason behind the diagnostic characteristic of disease is simple: thanks to science, we can reduce the risk of further illness, and possibly alleviate the existing discomfort based on an observation.

The WHO report on osteoporosis gave birth to a disease by describing a limit for bone density that could be used to diagnose. On one side of this line was bone loss considered typical of normal ageing, and with it came generic advice, such as exercising and eating more cheese and yoghurt. On the other side there was a heightened risk of fracture, and therefore a pressing need to take some sort of remedial action.

Despite the immense success of diagnostics, it's important also to realise that a diagnostic philosophy isn't without its own set of problems.

Herpes is a virus that fills many with disgust, especially if it's the variety that gives you blisters 'down there'. The same virus is also responsible for cold sores 'up there', around the lips. Either way, the WHO estimates that around 67 per cent of the population under the age of fifty has the virus.[36] That's around 3.7 billion people worldwide. Unfortunately, pathology results based on the antibodies in blood tests can't say whether you've come into contact with the virus and defeated it (leaving behind evidence of the Great Herpes War in the form of antibodies), or if you still have the virus. Then there's the issue of false positives for low readings. A swab of a blister or detecting proteins is more accurate, but in any case, a positive screen for herpes is a stigmatising and demoralising diagnosis that screams 'I have an STI', regardless of whether they've presented any symptoms. Sometimes a bunch of numbers that comes back from the pathology lab doesn't provide the full story of disease; a responsible general practitioner must interpret the results for the patient.

At the other end of the spectrum, in the absence of something physical to measure or a clear diagnostic result, it's easy for a physician to conclude that nothing is wrong.

In November 1990 the magazine *Newsweek* published an article titled 'Chronic Fatigue Syndrome'.[37] It wasn't the first piece to address this emerging condition, but it contained a quote by a physician named David Bell, which summed up an all-too-common attitude towards the illness: 'He gathered that the Nevada resort town had been hit by *Yuppie flu*, a fashionable form of hypochondria, and that two local doctors were blaming it on the Epstein-Barr virus. "I knew that whatever we were studying had nothing to do with Yuppies," he says.'

Dr Bell might have known that this new condition wasn't an invention of young urban professionals, but for many the emerging condition known as CFS was little more than the hypochondria of coddled twenty-somethings, a fabricated illness which privileged rich kids claimed to have when they were faced with hard work. Even as the science behind CFS progressed, conservative media such as the *Daily Mail* continued to display headlines such as 'Doctors are ordered to take "Yuppie Flu" seriously', as if their audience wouldn't recognise the condition by any other name.[38]

As recently as 2012 a British *Sun* newspaper columnist opened an article about welfare cheats by proposing that for a New Year's resolution he might fake a disability. 'Nothing too serious,' Rod Liddle wrote, 'maybe nothing more than a bad back or one of those newly invented illnesses that makes you a bit peaky for decades – fibromyalgia, or ME.'[39]

Yuppie flu, Chronic Fatigue Syndrome, fibromyalgia, myalgic encephalomyelitis (ME) – the condition (or conditions, depending on where borders are drawn) has been a revolving door of names over the decades. They all refer to a similar sense of exhaustion following exertion, complete with muscle aches and intense joint pain, unrefreshing sleep and a wide variety of other symptoms, from headaches to loss of appetite to anxiety and depression. The debate that's pushed back and forth since the 1980s makes for an interesting comparison with other historical diseases that seemed 'fashionable' for a short time before fading into obscurity. It's hard to avoid comparisons between CFS and neurasthenia.

One important difference separating the conditions, however, is a century of medical science. Much had been learned about the human body across 100 years, and by the late twentieth century identifying the biological roots of suffering was half the battle. Linking symptoms with biochemistry, microbiology or anatomy can provide us with vital clues on how to treat them. By the same token, though, when deprived of that smoking gun we're left with an uneasy sense of doubt that there's anything to treat, coupled with a fear that we might be unintentionally fabricating a disease. Regardless of the suffering a person might express, a disease just isn't a disease until we can identify a physical flaw.

With a rise in reported cases of fatigue accompanied by aches and general malaise, the hunt was on for an absence (or glut) of some hormone, a faulty gene or a rogue virus. Researchers first made a connection between these symptoms and an unusual epidemic of what was reported as poliomyelitis in Los Angeles in 1934. Unlike past outbreaks, these cases were non-fatal, with just a few incidences of persistent paralysis. The term *epidemic neuromyasthenia* was soon being used to describe the outbreak. Following a similar spike in infections in London around twenty years later, the term *benign myalgic encephalomyelitis* was used to describe the neurological symptoms and muscular pains that seemed to follow such viral infections. The diagnosis was added to the *ICD* under 'Diseases of the Nervous System' in 1969, making ME official.

Not that everybody was convinced. Many doctors saw ME as more of a psychosocial illness, since any physical signs on which they could base a diagnosis were varied and inconsistent. Some suggested it should be called *myalgia nervosa*, making it a purely psychological condition. Code for 'it's all in the mind', in other words.

Given that the condition affects a disproportionate number of women, a minority of physicians in the 1970s even likened ME to a form of hysteria. This association with femininity drew further comparisons between CFS and neurasthenia, both fashionable sicknesses coloured by our expectations of what it means to be a man or woman.

A scene in a 1999 episode of *The Simpsons*, 'Guess Who's Coming to Criticise Dinner?', presented a stereotypical perception of the disease better than any medical journal could. An editor lists for Homer the kinds of items readers expect in a lifestyle section – ending with Chronic Fatigue Syndrome – before nonchalantly labelling it 'chick crap'. Boom-tish.

In the 1980s, an outbreak of what was suspected to be glandular fever (or mononucleosis) in the US state of Nevada drew attention for again giving rise to chronic fatigue, inspiring the new name Epstein–Barr Virus Syndrome, after the pathogen responsible. In 1985 the Centers for Disease Control was tasked with nailing down a diagnosis. If the Epstein–Barr virus was responsible, they couldn't detect it. Not that it made much difference – the association between glandular fever and ME stuck firm, despite the lack of proof. There were subtle signs of scepticism in the literature, as a handful of researchers continued to champion social or mental explanations based on weak correlations they'd found between the symptoms and certain psychological tests.

The name Chronic Fatigue Syndrome was put forward in 1988, replacing Epstein–Barr Virus Syndrome as researchers finally came to a consensus that tests for the Epstein–Barr virus were useless as a diagnostic tool. Still, physicians were without a 'gold standard' method for diagnosing the syndrome.

In the ensuing decades, CFS has continued to be the focus of heated public debate and intense medical investigation. The symptoms have been linked with even more infections, such as Human T-lymphotropic (Type 2) and xenotropic murine leukaemia virus–related viruses, while other studies have once again refuted such correlations. The most up-to-date research points the finger at changes in gut bacteria that could be affecting the body's immune system.[40] Changes at a sub-cellular level, inside the power-providing mitochondria, indicate a reason behind the fatigue itself.[41] The puzzle pieces are gradually emerging, revealing a complex interplay of microbes, a rogue immune system, and cells changing how they deal with their energy needs on a deep, biochemical level.

With all that we're learning, the debate is far from over. Even the name itself has invited heated arguments, with some feeling the term 'chronic fatigue' fails to depict the magnitude of suffering, while others believe the history of myalgic encephalomyelitis doesn't accurately describe their personal experience of the symptoms.[Δ] A 2015 US Institute of Medicine report suggested the name Systemic Exertion Intolerance Disease, but that failed to win many fans.

Whatever we name the condition or determine its primary cause to be, CFS is another great case study in the defining of disease. Reported suffering isn't enough on its own – invisible symptoms require a visible variation in biology, one the afflicted has no control over (such as a virus or bacterium), a change in biochemistry or an unusual concentration of cells or hormones. Only then can the discomfort, fatigue and numerous other emotions, sensations and impediments be legitimised as a disease. And only then can the condition attract sympathy, research dollars and an earnest search for treatment. Without the designation of *disease*, there is an expectation that the sufferer should have a greater degree of agency over their symptoms. They are to blame: they are simply being lazy.

When a condition is thought to be 'all in the mind', there's the implication that the mind is controllable, and therefore the condition can be fixed by will alone. We use the word *psychosomatic* to describe physical experiences that seem to have no clear link to our anatomy. The term should carry medical weight, but nonetheless still implies a minor ailment, one that willpower alone should be

---

Δ   While I respect that there is an ongoing debate concerning whether they are the same illness or two distinct conditions, I've chosen to primarily use the term 'CFS' here for the sake of brevity and simplicity.

sufficient to cure. To go one step further, it's the disposition of the sufferer – their mental fortitude, their personality and even their morality – that's flawed.

As a consequence, sufferers of CFS have often lost trust in medical authority. The ME Association in the United Kingdom reported an online survey which found that just under half of respondents experienced 'poor' or 'dreadful' management of their condition by their general practitioner.[42]

Vast differences in how data is collected makes it hard to pin down a total number of CFS sufferers.[43] The best estimates suggest around fifteen out of every 100,000 people worldwide, although the trend seems to be heading south, with more people being given a diagnosis of fibromyalgia instead.[44] Exactly how much is down to changes in testing criteria, reporting or epidemiology is anybody's guess.

The binary model we use to separate disease from good health has, for the most part, been incredibly useful for reducing suffering in the world. If we stand far enough back, it becomes obvious that biology plays a lead role in causing us pain and misery. By understanding the complex system of chemical gears and hormonal pulleys that make up the human body, we have a good shot at alleviating discomfort and improving wellbeing. That kind of success isn't up for debate.

Nevertheless, the logic of disease is still based on a premise of relative normality. The culture surrounding an illness establishes a form of authority we all trust, one that determines who is broken and who is a pariah, who is forgiven and who is damned, which conditions deserve to be studied and which aren't worth our attention. That said, the success of medicine shouldn't be an excuse to turn a blind eye to its failures.

'Is CFS a disease?' is a much sexier headline than 'How can we help reduce suffering in people who are constantly exhausted and in pain if we don't know what's causing it?' But the question itself is fundamentally damaging. In asking whether suffering constitutes a disease, we presume that a condition is *either* a failure of morality *or* a failure of nature. In one direction lies the stigma of weak character; in the other lies the stigma of the biologically unfit. And in the middle is a person who still feels unwell.

# DISEASED

Let me introduce you to Jack.

Jack is the name we'll give a little bugger in a comprehensive school I taught at in London's east. Without fail, Jack would arrive late to class, start off-topic conversations, ask ridiculous questions, provoke arguments, and nine times out of ten end up outside the head teacher's office. He was rude, obnoxious and a risk to student and teacher safety. I was advised it wasn't safe to do certain science experiments with Jack in the room, so by order of my head of department it was only when Jack was on suspension that we could bust out the Bunsen burners and concentrated acids. That wasn't often, until at long last he was booted out of school after one too many punch-ups. So long, Jack.

But Jack had dyslexia, you see, and anxiety. According to his individualised education program, he also had attention deficit and hyperactivity disorder, and there was an item in Jack's record mentioning something about a history of oppositional defiance disorder. There was a buzz around the school that his learning difficulties stemmed from foetal alcohol syndrome or some other drug-related disorders, not that I ever confirmed that.

I had a soft spot for kids like Jack. If you ever met his parents, you'd quickly start wondering how he coped with anything at all.

'In my day there was none of this ADHD bullshit,' his father griped in a meeting one afternoon. 'You'd just give 'em a clip around the ear – didn't do me any 'arm.'

In another time and place, Jack was a bad kid who needed a beating to 'set him straight'. In my classroom he was hobbled by his biology, requiring patience and assistance. Out on the football field he was a star player who might one day play for his favourite team, if he put in the work. In an academic environment he couldn't sit still and read three pages of 'Chapter 3 – Ecosystems and You' without throwing a tantrum. Or the textbook.

The question of what turns the right mix of nature, nurture, predestination and agency into a disease is one for the anthropologists. While many, if not most, pathological conditions are more or less consistent across countries and centuries, different ethnic communities also have their own unique takes on what makes something a disorder, as opposed to just a different way of being human.

Every society throughout human history has had Jacks by the hundred. It's the job of the medical anthropologists to study them. One way they tend to determine whether a person like Jack is sick or sinful is by comparing their observations with three criteria. So let's start with question one.

*1 Is Jack different?*
Just like your mother used to tell you, you're special. What she didn't tell you is that everybody is. Including Jack. So you're both as special as any other life form, from an amoeba to a zebra. How's that for depressing?

The evolution of life depends on variation. Small differences in our genes are responsible for significant differences in our biochemistry. Genes interact with our unique environments to create differences in size, shape, chemical constitution and function. You share about 99.9 per cent of your genetic code with every other human, making all of us incredibly similar, at least compared to many other species. Yet that 0.1 per cent of variation can result in a nearly infinite range of tiny functional differences.

We all have unique communities of bacteria in our digestive systems, and a zoo of parasites and fungi on our skin. Latent virus codes lie buried in our DNA. The neurons that make up our nervous system are connected in infinitely unique ways, with synapses pumping out slightly different concentrations of neurotransmitter, bestowing each of us with different memories and thoughts and temperaments. Our metabolisms are all slightly different, distribute fat in different areas and affect how we feel at different times of the day.

Some of those differences are so common that we give them little thought. Blue eyes, tans, red hair, aquiline noses – they're all acceptable variations. A percentage of people periodically lose large amounts of blood. Ghastly, maybe, but menstruation isn't normally thought of as a disease, even when it's accompanied by debilitating cramps, pain, chronic lethargy or emotional change. In fact, we all *expect* to be different in certain ways. A receding hairline and lost collagen are part and parcel of getting old. But if you experience this at age five, there's a chance you could be diagnosed with a disease called progeria.

Since Jack behaved in a way that lies beyond our expectations of typical variations, it's possible he could have a disease. We expect our bodies – and those of others in our society – to look, sense and perform in certain ways. There is a spectrum of variation which we tolerate, such as where hair grows, the number of moles and blemishes, certain speech inflections, an amount of discomfort and lethargy, and how well we can see and hear. We also tolerate minor variations in how much patience, empathy, respect and resilience we have. Our expectations are influenced by the people we commonly associate with, and by the nature of the stories we tell that celebrate our forms and functions.

But simply being different is not enough for us to say that Jack's behaviour is caused by a disease.

*2 Okay, so Jack is different. But is Jack deficient?*
Even when it came to starting a war, the Prussian statesman Otto von Bismarck understood the significance of etiquette: 'Be polite; write diplomatically; even in a declaration of war one observes the rules of politeness.'

In every society, individuals are expected to meet certain social standards and obligations. First and foremost, we're all expected to make others feel comfortable in how we act and present. Don't talk to yourself out loud on public transport. Don't go out looking too unhygienic or too ugly. Don't act in an aggressive or violent manner. There are standards we must meet so as not to cause alarm or discomfort in others.

Secondly, we must provide for ourselves and support others around us. In capitalist societies, this usually means we have to work for an income, which we then exchange for goods and services. In non-capitalist cultures, this might manifest as an obligation to collect food, provide comfort or memorise important stories or laws.

We are expected to make use of the available resources and infrastructure, to not take too much from others, and to contribute to the birthing and raising of a generation of children. While we can expect some aches and pains in life, a certain amount of discomfort becomes debilitating. When a variation in biology prevents us from meeting our customary responsibilities, a mere unwanted difference becomes a potential disease.

Lacking perfect vision isn't necessarily a disability. Even those who need glasses when they read a book in the evening would not necessarily think of themselves as having a disorder or disability. Only when their abilities to fulfil society's expectations are compromised – such as reading street signs while driving – does bad vision become something more pathological.

Feeling tired isn't a disease. Regularly waking up so exhausted you can't possibly work, on the other hand, indicates a disorder. Having a bump on your nose or a small birthmark isn't a disease. Being covered with bumps or patchy discolourations – even if benign and painless – just might be.

So we can say that diseases are abnormal variations in biology that interfere with our expectations of what it means to be happy, comfortable and responsible. We expect adolescents to conscientiously devote time to learning skills that will benefit society and themselves; if Jack's brain prevents him from doing that, we might say he has a disorder.

Yet if the concept of disease is just about differences in biology that stop us from meeting our social responsibilities, then what should we make of hangovers? After all, alcohol toxins can wreak havoc with our circulatory system and gut, and can definitely keep us from heading into the office for at least half a day. Most humans are usually fairly sober, so hangovers count as a variation in biology that stops us from fulfilling our obligations. If those same symptoms were the result of a bacterial toxin, few would question it being a disease such as food poisoning.

So to regard something as a disease, we need to specify one final important qualification.

*3 Right, so Jack is different, and he's certainly deficient. But is it his fault?*
Arguably the most significant factor separating diseases and disorders is *culpability*. With disease, nature is to blame. It's the random infection of a passing virus, the chance breaking of a cancer-causing gene, the toxic build-up of heavy metals in your food and water. With it comes sympathy, if not forgiveness for not fitting in.

In the words of the medical anthropologist Allan Young: 'For the anthropologist, sickness can be conceptualized most usefully as a kind of behavior which would be socially unacceptable (because it involves withdrawal or threatened withdrawal from customary responsibilities) if it were not that some means of exculpation is always provided.'[45] In many cultures past and present, this lack of culpability has taken the form of spiritual affliction. You aren't to blame for your weakness, pain or violence because some supernatural being or agent of nature has taken over, assuming responsibility on your behalf.

Even in more 'enlightened' cultures, agency impacts heavily on how we categorise and respond to diseases. Body modification isn't a disease, no matter how taboo the scars, tattoos or implants. The side-effects of illicit drugs or alcohol aren't diseases, no matter how much they make your stomach heave or head spin. If you decide to stay in bed all day, you don't have a mental health disorder – you're just lazy. If you choose to eat too much and grow morbidly fat, you're just gluttonous.

There are some counter-examples, of course. It's possible to argue that tinnitus is still a hearing disorder, even if you chose to listen to loud music without earplugs as a teenager. If you get chlamydia through unprotected sex you are still catching a disease; lung cancer acquired through smoking is still categorised as cancer. Yet even if they qualify as diseases, we don't treat all conditions the same when they're associated with 'vice' or lifestyle choices.

In the United Kingdom, lung cancer accounts for 22 per cent of all cancer deaths but only gets about 8 per cent of National Cancer Research Institute funding. By contrast, breast cancer accounts for just under 8 per cent of deaths but gets about a quarter of the funding. Leukaemia gets about 20 per cent of the funding but accounts for only about 3 per cent of deaths. Breast cancer and leukaemia awareness campaigns are popular causes; the sense that smokers somehow deserve their fate creates a greater challenge for lung cancer research fundraisers.[46]

When HIV became an epidemic in the 1980s, its stigma as a disease of homosexuals and drug addicts saw it receive far less public demand for a cure than it would have if it struck individuals at random. At a White House press conference in October 1982, by which time there had been some 600 cases of AIDS, a journalist named Lester Kinsolving grilled deputy press secretary Larry Speakes on whether the Reagan administration knew anything about the epidemic.

'Over a third of them have died,' Kinsolving explained, after Speakes had admitted the White House didn't have anything to say on the matter. 'It's known as "gay plague".'

The press pool laughed. What followed wasn't concern, or a promise to follow up on the condition with the president, or even quiet commiseration.

'I don't have it ... do you?' Speakes asked the journalist.

It was the first of a number of light-hearted and homophobic remarks. The White House statements reflected an attitude that would persist for years, framing AIDS as a gay epidemic and therefore not an illness worthy of investment. Only when HIV found its way into blood transfusions – an unnecessary and tragic additional cost of their hesitation – and therefore into individuals who had not made choices viewed as morally corrupt, such as those with haemophilia, was the virus given serious attention.

As with addiction and obesity, slapping on a label of 'disease' doesn't change the fact that culpability plays a strong role in how we fund, treat and talk about health conditions that are associated with morality or willpower. Jack isn't sick if he's considered to have some control over his actions; biology be damned. But absolving Jack of guilt comes at a cost: declaring a state of health as a disease can remove his sense of responsibility for his own actions, and even his agency over his body altogether.

A rather colourful and controversial figure in the history of psychiatry, Thomas Szasz, published a book in 1961 called *The Myth of Mental Illness*.[47] Although he was a psychiatrist, Szasz was no great fan (to put it mildly) of his profession's culture. He saw mental illness not as an organic condition in the way cancer or malaria are, but as a disorder purely of the mind.

Szasz argued against the belief that mental illnesses were diseases of the brain in order to remove stigma and fight the authority of the state institutions that controlled those suffering poor mental health. In his own words, he set out to 'undermine the moral legitimacy of the alliance of psychiatry and the state'.

To do so, Szasz had to define *disease* as something material – an alteration of normal, physical body structures. As physical problems, 'such as a misshapen extremity, ulcerated skin, or fracture or wound', medical diagnoses could be supported or falsified using 'more specialised methods of examining bodily tissues and fluids'. Since we couldn't possibly do that with mental health problems, Szasz decided, labelling them as diseases was wrong. 'Mental illness, of course, is not literally a "thing" – or physical object – and hence it can "exist" only in the same sort of way in which other theoretical concepts exist,' he claimed.

On the one hand, it's not hard to sympathise with Szasz's basic position. If history is anything to go by, when someone's state of mind is declared 'diseased', this is often used to justify removing their autonomy – and the 'treatment' they are administered can result in terrible violence.

A pair of diseases 'discovered' by the American physician Samuel Adolphus Cartwright in the antebellum United States illustrates how medicalising a behaviour can legitimise the taking of control against a person's will. In his book *Diseases and Peculiarities of the Negro Race*,[48] Cartwright described a condition 'peculiar to negroes' and 'well known to our planters and overseers' called drapetomania – a disease that caused African-American slaves to seek freedom, against God's will. The cause, Cartwright believed, was masters who treated their slaves as equals. In addition to this condition, Cartwright also proposed that laziness among slaves was the result of a malady called dysaesthesia aethiopica, which was caused by an insensitivity of the skin. It could be treated by having 'the patient well washed with warm water and soap'; then you rub them with oil and 'slap the oil in with a broad leather strap; then ... put the patient to some hard kind of work in the sunshine'.

Cartwright supported his claim that these were legitimate diseases on the grounds that they were in conflict with how God created mankind. By framing errant behaviours as diseases caused by lax or sympathetic slave owners, Cartwright appealed to the natural law of biology rather than morality of humans, reducing slaves to broken machinery that could be fixed, but only if their owners were granted absolute authority.

Szasz probably wouldn't have seen much difference between Cartwright's inventions and many of the mental illnesses in his own time; likewise, it's tempting to draw parallels between how Cartwright imposed diseases on slaves in order to 'treat' them and how we might impose a disease such as oppositional defiance disorder or attention deficit disorder on Jack, only to 'tame' it with medication or incarceration.

Advances in technology have steadily added to what we know about how the brain works, allowing us to better understand differences in neurochemistry and neural pathways, which can account for variations in human behaviour. So while it's problematic to think of our behaviours and mental functions as 'broken' or 'functional', it's equally troublesome to divorce them from the physical brain, or to believe that some of these deviations don't cause discomfort in individuals.

Szasz's fight against mental illness as a disease is an example of how our very thinking about disease has become tattered. We're confused over where organic flaws end and personality begins. The question of whether something is a disease polarises health into 'normal, and therefore should be outcast' and 'broken, and therefore should be treated by somebody who knows better'; there's little room in between. Tragically, getting it wrong can have dire consequences. Oversimplifying the causes of poor health and disease has in fact been shown to hinder our acceptance of people suffering from mental health conditions.[49]

Poor Jack is broken. To work out what kind of broken, and then to determine the best way to help him, we study his physiology and question his agency, and then place these on a set of scales and make a judgement. That is how we diagnose Jack's disease, and we've become quite good at it.

# II
# INFECTED

# INVADED

Degrees in microbiology teach you that our world is positively loaded with germs. So they create two types of graduate: those who are increasingly paranoid, and those – like me – who are suddenly aware that they've *always* been surrounded by germs, and become a little more complacent.

Not being the paranoid type, I'll happily eat food after dropping it on my kitchen floor. (Five-second rule? Try twenty!) I flout use-by dates, and even eat around the burgeoning spots of mould in the last piece of bread, moderately secure in the knowledge that so long as I respect the basic rules of good hygiene, there's little chance I'll land in the emergency department in need of several units of IV saline.

I mightn't be put off by the odd bit of muck, but I'm not blind to the less friendly microbes that lurk in our midst. I've felt the scorching-hot skin of an infant who has a soaring fever. I'll never forget the necrotic stink of runaway ulcers on the limbs of patients suffering diabetes. And whenever I slipped a drop of blood into a diagnostic machine or smeared one onto a slide, the thought of contracting a life-changing virus was never far from my mind.

For something so tiny, and until recently so hidden, the discovery of the microscopic life form revolutionised how we distinguish clean from dirty, and healthy from diseased. As a result, we've become veritable Vikings when it comes to sterilising our world.

In a search for the definition of disease, it's hard to not look to the living contagion. The very concept of disease and cure has become almost synonymous with the scourge of the germ. Disease is an invasive enemy we divide into discrete categories, and medicine is the bullet that eradicates disease. Philosophers label this concept of invasion the *ontological model* of sickness. Leaving aside the fancy words, it views disease in terms of a 'thing' we catch. Beneath the disease, therefore, there's still a healthy, normal person just waiting to return to vitality.

Most of our language and behaviour around disease today is ontological. And we're not just talking about invading microbes. The 'C word' offers a perfect example.

'Cancer isn't something you fight,' Liz explained as we waited for our hot chocolates to arrive. 'Fuck that.'

Liz swore a lot, and so eloquently that it wouldn't sound out of place in a parliamentary speech. I loved that about her. And she had strong opinions about cancer, which she expressed passionately in her blogging and tweeting. After all, it would soon take her life from her.

Early in 2016, in a small cafe across from Newtown Station in Sydney, Liz and I exchanged our last words. A short time later I would be in the United States, too far away to return when I finally received word that she'd passed away. Truth be told, I'm not much good at goodbyes anyway. Our departing 'see you later' still hangs unfulfilled.

From the day we met after her doctor's appointment to that warm autumn afternoon in Newtown, we must have discussed death and cancer a hundred times, over sushi and scotch and rides home from evenings at the cinema. Often clinically, sometimes personally, always with expletives.

'There is nothing more fucking passive than being sick,' she said, her face drawn but her eyes still full of the passion I adored in her. 'You don't fight it. You can't. You lay back and let science do the work. You then get better or you die.'

Liz rebelled hard against the misguided, if well-intentioned, advice which others imposed — of herbal remedies or detoxifying soups or mantras of positive thinking. She was a champion for medicine, even if she understood the temptation to pursue alternative treatments. It wasn't hard to sympathise with the desire for a miracle. Even a doomed strategy was an action, she once said, and acting felt far more empowering than lying there as a prescribed dose of heavy-duty chemicals dripped into your vein.

Liz's cancer wasn't ontological. Not to her. It wasn't some demon that could be exorcised or a toxin that could be expelled. It was just shitty genes.

She hated being called a hero. So don't. And while Liz knew she could squeeze valuable days out of her life with chemo and surgery, she also knew that, ultimately, her body would one day fail. It wasn't an outside force attacking her body. It was just an unfortunate fact of her biology.

In contrast to the ontological concept of disease is the *physiological model*. It holds that a disease is still 'you', but not in a way you might like. This picture of disease can't be considered an invading army you can fight – it's your own recalcitrant nation, rioting in all-out civil war.

These two contrasting ways of thinking about disease have each had their moment in the sun in different parts of the world and at different times in history. In the ancient West and the Near East, disease was often described as physiological. Go back to classical Greece, for example, and you'd learn that health was all about levels of liquids in the body called humours. Just four fluids – blood, yellow bile, black bile and phlegm – made up everything from sweat to puke to semen to spinal fluid. If these went out of whack, an imbalance called dyscrasia, your body let you know it.

Hot and sweaty from fever? Your body has too much of the warm, salty kind of liquid. Finding ways to help the body purge itself of (or replenish) these four substances – through resting, vomiting, defecating, abstaining from sex, having more sex, sweating, drinking, not drinking, or eating the right foods – was the way to let the body find its balance again. If it didn't find balance, it would reach a crisis point, where all bets were off. That was when you'd better write out your last will and testament.

Thanks to an obsession with ancient Mediterranean wisdom, the notion of balancing humours persisted through the Middle Ages well into the nineteenth century. Bloodletting and blistering, along with prescribing emetics, laxatives and diuretics, were still occasionally performed as treatments in the belief that sickness was somehow related to too much or too little fluid.

In spite of its popularity over the centuries, this model of sickness wasn't all that good at saving lives. It certainly wasn't as successful as the radical new theory of germs that slowly but steadily took the world by storm. That's kind of ironic: becoming more 'civilised' might have brought us medical revolutions, as well as the printing press, cheap takeaway food and the PlayStation, but it also brought us continent-crossing plagues, pestilences and poxes to begin with. Piling humans in close proximity to one another, along with their food and faeces, sets up the perfect conditions in which infectious diseases spread.

Statistics of the regular, far-reaching plagues through the last millennium make for shocking pub-trivia facts. Up to 60 per cent of Europe's population died in the fourteenth-century Black Plague, with as many as 90 per cent in some regions.[1] During Napoleon's retreat from Moscow in 1812, more of his soldiers died from typhus than by Russian attack.[2] The 'Spanish flu' killed about 50 million people between 1918 and 1919, after infecting about a third of the world's population.[3]

Pinpointing the causes of pestilential outbreaks hasn't always been easy, given the microscopic size of the agents responsible. Considering that infected victims typically shared the same damp, unhygienic stench amid flying vectors such as mosquitos and flies, it's not hard to see why the air itself was mistakenly blamed for illness for so long.

Medical writers schooled in the philosophies of that most famous of ancient Greek medics, Hippocrates of Kos, blamed filthy conditions for 'stains' on our health as far back as the fourth century BCE.[4] A few centuries later, the Greco-Roman super-physician Galen explained that bad air caused an imbalance in the humours, while the Roman architect Vitruvius warned of the dangers of 'noxious exhalations' from the marshy districts or unwholesome quarters in his famous text *Ten Books on Architecture*.[5]

It wasn't until the 1300s that writers got into the nitty-gritty of discussing plagues, with numerous essays explaining the treatments and causes of all-too-common epidemics. The summary version of these essays goes like this: putrefied air was responsible for spreading illness, having been belched up from marshes and swamps, blown in from some far-off land, or occasionally released from beneath the ground by earthquakes. The fix was simple: just make the stinky, corrupted air smell better with incense or perfumes. Meanwhile, civil planners needed to build better sewers and drain local swamps. A fourteenth-century British parliamentary statute made it an offence to dump 'dung, offal, entrails, and other ordure into ditches, rivers, waters, or other places', given that these corrupted the air and led to disease. The fine was £20 – enough to have bought a decent horse or two.[6]

Sometime before the nineteenth century, academics borrowed a Latinised Greek word for pollution – *miasma* – to describe this foetid atmosphere. Soon the toxic particulates responsible for diseases such as dysentery, tuberculosis and cholera were being called 'miasmata'. These weren't germs in the sense we'd think of them today, but the concept did serve as something of a stepping stone between disease theories.

Miasma models, in all their forms, pitted corrupting forces against pure, normal functioning. The idea wasn't merely a placeholder for our modern belief in pathogenic bacteria, protozoans and viruses. Pathogens, we now know, are substances that invade our bodies, grow in number and then redistribute. Instead, miasmata were noxious elements that corrupted health, much like a spray of salt water eats away at iron. An early-nineteenth-century English physician, Thomas Southwood Smith, warned in the General Board of Health's *Report on Quarantine* not to 'overlook ... the corruption of the air', in the face of a growing belief that disease might rather be spread by touch.[7]

The idea that sickness could be spread by physical contact wasn't exactly novel in the 1800s. Three centuries earlier, a physician and part-time poet by the name of Girolamo Fracastoro was an early advocate of not just contagion theory, but cleanliness in general. We can also thank him for inspiring a change of name for a common venereal disease in a 1530 poem.[8] In it, a mythical shepherd named Syphilis curses his god over a drought and receives the disease in punishment.

According to Fracastoro, there were a few kinds of contagion, all of which acted like seeds, entering a body and blooming into illness. The poet didn't describe the seeds as alive, but rather as a material that could evaporate and diffuse through the atmosphere, cling to clothing or spread by touch. Tiny things must have fascinated the man, as he was also an early fan of atomic theory. Talk about ahead of his time!

It took advances in the microscope during the seventeenth century to open the public's eyes to the living contagion. For a while it was fashionable among the well-to-do to put muddy water and tooth plaque under the lens as a form of pre-dinner entertainment. More serious scientists, with the exuberance of modern adolescents, would check out their own semen and dandruff. The Dutch fabric tradesman Antonie van Leeuwenhoek took a break from using his microscope to analyse fabric threads to observe muscle tissue, sperm and pond water. He was the first to describe the presence of a zoo of what he called 'animalcules' – tiny moving creatures, too small for humans to see with the unaided eye.

A few imaginative scholars speculated that such tiny critters could be the cause of a number of illnesses. The polymath Jesuit and general overachiever Athanasius Kircher used a microscope to study the blood of plague victims in Rome, and wrote in a 1658 treatise of 'little worms' that could be seen in putrefying matter.[9] He wasn't alone. In 1700 the French physician Nicolas Andry wrote in his book *De la génération des vers dans les corps de l'homme* (translated into English as 'An Account of the Breeding of Worms in Human Bodies') that smallpox was caused by similar microscopic worms.[10]

Given the quality of their equipment – and the fact that smallpox is caused by particles of viruses that could not have been visible under their lenses – it's unlikely that these were early observations of the pathogens responsible for these diseases. It's far more likely that they were seeing chains of garden-variety bacteria. Still, these reasoned guesses foreshadowed the debate over the causes of disease, a controversy that would take centuries to resolve.

Why so long? As novel as microscopes were, the technology took a long time to become powerful enough to study the nature of these itty-bitty sticks, blobs and balls. Before long, microscopes became passé. A French physiologist named Louis Mandl remarked in the mid-nineteenth century:

> [T]owards the end of the last century, the microscope experienced the fate of so many other new things; having exaggerated its usefulness and used it to support lunatic flights of fancy, people went to the other extreme and exaggerated its inconveniences and hazards; then its use was almost completely neglected, and results obtained with it were only spoken of with mistrust. Even the existence of blood corpuscles was doubted, and what Leeuwenhoek and his successors had described were attributed to optical illusions.[11]

Meanwhile, fascination in 'tiny animals' among academics waned. It wasn't until the mid-nineteenth century that microscopes finally allowed us to make out details in the life cycles of organisms the size of bacteria. With new and improved tools, the medical establishment could finally explore the possibility that some diseases were caused by near-invisible, reproducing materials.

The German anatomist Friedrich Gustav Jakob Henle was convinced that malaria was caused by miasma. Smallpox, on the other hand, had to spread through direct contact with other humans. As did measles and influenza. It stood to reason that if a disease-causing agent was in one person, for it to spread to two or more people it had to somehow expand in mass. Poisons, Henle figured, couldn't spontaneously increase in substance, so there had to be some other explanation. In his treatise *On Miasmata and Contagia*, he referred to these agents as *contagia animata*, and discussed ways one might prove definitively that a given disease was caused by a reproducing particle.

Old ways of thinking are hard to shake. Notions of balance and corruption continue to inform health decisions today. The shift from one disease model to the next is imperfect. But in some ways, the revolution couldn't come soon enough – medicine was murderous when it consisted of purging, sweating and unsanitary cutting. Even for a germophile like me, history is a horror show of pathogens, one that steadily grew worse as cities swelled and populations multiplied.

Luckily, a revolution was exactly what the world got.

# CONTAMINATED

We don't have an international day to celebrate Jakob Kolletschka's death. Maybe we should. A pivotal moment in the whole debate between bad air and contagions took place in 1847, thanks to the fact that this Bohemian doctor cut his finger, got sick and died.

One reason we don't celebrate it could be that his friend was something of a belligerent asshole. Not that I really blame the guy. But if it weren't for a toxic mix of nationalism and arrogance, it's possible the evidence provided by Kolletschka's Hungarian friend might have landed a solid punch for germ theory a decade or so earlier than fate had it.

Let me tell you about an obstetrician from Budapest by the name of Ignaz Semmelweis. He and Kolletschka worked in the maternity wards of the Vienna General Hospital, observing patients, assisting midwives with complicated births and teaching medical students the more complicated side of how babies are made. Semmelweis was also responsible for record keeping across the hospital's two obstetrical clinics, keeping note of patient admissions, births and deaths.

Maternity institutions such as the one at the Vienna General were effectively charities established to counter high rates of infanticide across Europe, especially among local sex workers. Women in poverty took advantage of the free medical assistance, and in return the hospital took responsibility for any unwanted children. Yet it was a widely known secret among the city's lower classes that if you were due to give birth, you'd best pay close attention to which day it was.

Giving birth is risky business for a mother at the best of times. Delivering your baby in Vienna General Hospital's obstetric clinics added the risk of dying from the dreaded puerperal fever – a nasty streptococcal infection that inflamed the uterus after childbirth. It was common knowledge that the First Obstetrics Clinic had a much higher death rate than the nearby Second Obstetrics Clinic, a fact Semmelweis knew all too well. Each clinic took in patients on alternating days, making the chance of getting the 'good' clinic a flip of the coin for expecting mothers. It wasn't unknown for women to delay attending the clinic – some even gave birth outside on the street – so they could avoid the curse of the First Clinic while still legally handing them the newborn.

Their fear wasn't blind panic spurred by overzealous rumours – they really were safer delivering out in the gutters than in the hospital's wards. Mortality rates fluctuated wildly, but deaths from puerperal fever averaged around 10 per cent in the First Clinic. In the Second Clinic, and out on the streets, it was a steady 4 per cent. Semmelweis was not just perplexed by this contrast, but sickened. He claimed to feel so saddened that 'life seemed worthless'. The physician set out to determine the cause of these fevers, which were generally dismissed by doctors as a fact of hospital life. He scanned his records, comparing everything from patient density to temperature to religious affiliation, ruling out variables one by one.

While on a vacation away from the hospital, Semmelweis received word of Kolletschka's untimely death. During a routine dissection of a cadaver, the doctor had been nicked by a scalpel in the hands of a nervous student. It didn't escape Semmelweis's attention that the systemic infection that took his friend's life was remarkably similar in its symptoms to those that had claimed countless mothers and babies over the years.

One key difference between the two Vienna clinics was the presence of obstetricians with their classes of students in tow. While deliveries were left to the midwives in the Second Clinic, mothers in the First Clinic were regularly poked, probed and palpated by the grubby hands of a queue of inquisitive novices and their teachers. To make matters worse, it wasn't uncommon for lessons on cervical dilation to occur right after a practical session of cadaver cutting.

For Semmelweis, Kolletschka's death was a light-bulb moment. He saw a direct line connecting the corpses with the uterus, and figured that if some 'cadaverous particles' had contaminated his friend, it was likely that this deadly matter was infecting the mothers as well.

Strictly speaking, Semmelweis didn't see puerperal fever as *contagious* as much as *transmissible*. 'Only smallpox can produce smallpox, which is what is meant by contagious,' he said. 'But pus from an abscessed tooth or cancer can cause puerperal fever.'[12]

In fact, Semmelweis was not the first to make a connection between rotting bodies and sickness; a few years earlier, the British physician Edward Rigby told his peers: 'Where a practitioner has been engaged in the post-mortem examination of a case of puerperal fever, we do not hesitate to declare it highly unsafe for him to attend a case of labour for some days afterward.'[13] But Rigby's advice fell on deaf ears, his colleagues choosing instead to air out their rooms to fight the dreaded fever. Across the pond in Boston, an American physician named Oliver Wendell Holmes also noticed a relationship between disease and hygiene, and argued that anybody who ignored the advice to give one's hands a scrub was guilty not just of negligence but of a crime.

Semmelweis wasn't aware of the views of Rigby or Holmes, but he used the authority he had as a professor's assistant to prescribe for doctors and students a regime of handwashing with chlorinated water between post-mortem examinations and rounds through the maternity clinic. Within weeks, the death rate in the First Clinic declined, and soon matched that of the Second Clinic. The data was irrefutable, and he figured his cadaverous particle theory was in the bag. So three cheers for Dr Semmelweis, champion of contagion theory, curer of puerperal fever, and destroyer of miasmas … right?

Not exactly.

Remember, he wasn't the most well-liked chap in the business. Although the facts were plain for all to see, numbers alone rarely make an argument compelling when an established culture is firmly wedded to other ideas. Semmelweis was attacking a conservative medical tradition, in which doctors had no problem fingering a cervix with gore-stained digits. Worse still, he was an outsider in Vienna, visiting from Budapest in a time when upstart Hungarian students were protesting the empire's oppression.

Semmelweis's letters to prominent surgeons, in which he lambasted them as 'murderers', did little to endear his views to them either. Lastly, Semmelweis was no fan of writing academic papers, choosing to point at the tables as if the data spoke for itself, rather than write persuasive letters to fashionable scientific societies across Europe.

As Semmelweis was mailing out his colourful insults and demanding a medical revolution, across the channel in London a man by the name of John Snow was also using the power of statistics to convince his fellow physicians that contagions, not miasmas, were responsible for a serious outbreak of cholera.

Snow kick-started the modern field of epidemiology by tracing the illness among Soho residents back to their use of a single outdoor water pump. Already a fan of contagion theory, this was the evidence Snow needed to make his case. It wasn't quite a smoking gun: many residents had moved away from the offending neighbourhood as the epidemic worsened, making it hard to irrefutably conclude that there was a trend. But Snow used the data from his interviews to convince members of the local council to remove the pump's handle, putting an end to the 1854 Broad Street cholera outbreak. It turns out John Snow did know something after all.

Meanwhile, poor old Ignaz Semmelweis returned to Budapest a bitter man, his failure to force radical change leading him to a life of obsession and depression, of drink and debauchery. In 1865 he was referred to a Viennese asylum, lured there by the offer of a tour of a new medical institution. Semmelweis twigged at the last moment but was beaten into submission by the asylum's wardens. He was dragged inside. Days later the doctor succumbed to septicaemia, possibly as a result of the assault.

Science is impotent if it doesn't lead to action. Tables of numbers mean nothing if their reliability, accuracy and relevance cannot be communicated in a convincing fashion. Although his theory has been vindicated, Semmelweis's sad tale is a lesson in how scientific ideas evolve within a critical culture. The doctor wasn't without supporters, and his real-world experiment wasn't ignored by contemporary proponents of contagion theory. But slighted by ethnic bigotry, coupled with an arrogant belief that facts should speak for themselves – not to mention his acerbic communication style – it's no surprise that Semmelweis failed in his efforts to overturn a firmly established tradition.

Medicine has always progressed in fits and starts. At times it's affected less by numbers and more by cultural pride, personalities and ideologies. Fortunately, reality is more stubborn than fashion.

Over the following decade, Semmelweis's ideas about 'cadaverous particles' collided with the world of microscopic 'animalcules' and practices of sanitation, delivering novel methods that saved numerous lives and improved health and wellbeing worldwide. Contagion theory accomplished what a belief in miasmas couldn't. Despite its immense success in improving urban sanitation and domestic hygiene, ultimately miasma theory could not match the predictive power of contagion theory, and so help end suffering.

That was pretty much the end of the science of stinky, corrupting gases, and the beginning of spreading particles of doom. It was a subtle shift but an important one. Physicians and scientists started to twig to the discrete, core elements of disease that enter the body and change it, and gradually gave up their belief that disease emerged from within as a conflict or imbalance.

There were still a few more elements to add to this brew before the modern definition of disease was well and truly cemented. For contagions to become the pathogens we know and fear, the very definition of life itself needed to be stripped down to its core.

# FERMENTED

I home-brew. It might be a bit of a cliché, but there really is nothing quite as magical as throwing a packet of dried microbes into diluted mash and waiting for a delicious stout to bubble forth. Did I say magical? I meant cheap. And it takes more effort than just buying a six-pack. But to be a little more romantic about it, brewing connects my love of microbiology with an ancient tradition. You might like to breed rabbits or fish, but at least I get to enjoy the waste of the pets I breed.

Civilisation and *Saccharomyces cerevisiae* were old friends by the time a French chemist named Louis Pasteur turned his attention to an industrious single-celled fungus known as yeast. Although scientists had barely grasped the fact that yeast was a fungus by the 1850s, it was long known to play a necessary role in turning the sugars in grape juice and soggy grain into one of the world's oldest party drugs – alcohol. The only question was how.

The favoured theory of the day was put forward by the early-nineteenth-century German chemist Justus von Liebig, who argued that fermentation was a purely chemical process. In simple terms, sugar sometimes just crumbled into alcohol thanks to unstable 'affinities' in the carbohydrate molecule. Yeast might play a cameo role by kick-starting the process as it decayed, but von Liebig wasn't convinced the material was essential.

In 1856, when Pasteur was dean of sciences at Lille University, a student named Emile Bigo-Danel complained that his father regularly had trouble fermenting sugarbeet into alcohol. Sometimes he'd have low yields, and the fermentation would smell like sour milk.

Louis knew a thing or two about fermentation already, having earned his stripes proving that differences in the polarisation of light through crystals of tartaric acid was caused by mirror versions of the same molecule. He used a microscope to investigate Bigo-Danel's soured beetroot juices, and noticed that their yeast cells looked a little odd. In the fermented solution that contained alcohol, the cells were round globules. He discovered that the soured fermentation mixtures contained small, elongated cells, what Pasteur called 'lactic' yeast. (We now know they weren't yeast at all but bacteria.)

Following further experiments, Pasteur declared that yeast produced alcohol when they grew in an oxygen-free environment. Similarly, lactic acid was produced by yeast-like microscopic organisms when they were suffocated. To prevent spoilage caused by this growth in any solution we might want to consume, such as milk, Pasteur advised heating the substance just enough to kill off the microbes, but not so much as to ruin the food. So began the process we now call *pasteurisation*.

The discovery wasn't just a boon for lovers of milkshakes. It revealed that von Liebig's belief was all wrong: yeast in fact had a starring role in the fermentation process.

Louis might have a neat process named after him, but he wasn't the only French scientist who claimed to have made the connection between yeast and fermentation. Pierre Jacques Antoine Béchamp studied how different sugars changed into one another under various conditions. While he initially thought water alone was enough for these reactions to occur, a few years later he changed his mind and blamed the 'mould' that grew in solutions. Béchamp contested Pasteur's claim to being the first to link yeast growth with changes in chemistry, leading to a dispute that the two never resolved.

Regardless of who deserves the blue ribbon, the discovery was a watershed moment in biology. Chemicals such as alcohol and lactic acid were clearly products of microbial growth, and this discovery set the stage for a debate about the very nature of life itself. What makes living things so different to other things in the universe?

Mud, fire, rocks – these things aren't alive. Plants and fungi grow and reproduce. Animals also grow and reproduce, and move. People are mobile like animals, but also express rational thoughts.

Well, occasionally. For millennia these characteristics have perplexed notable philosophers. To the Greek philosopher Aristotle, there were different forms or essences (which, for lack of a better term, are often translated as 'souls' – not so much in a religious sense as in a nebulous *je ne sais quoi* sense): a vegetative one, which drove things to grow and reproduce; an animal one, which made things move; and a rational one, which inspired thought. Aristotle's theory of souls influenced centuries of theological debate on life, which considered anything from moss to Moses as the product of a discrete, non-material, supernatural force.

As late as the early to mid-twentieth century, this notion of a pervasive living spark – broadly defined as *vitalism* – stuck around as a serious scientific concept. It was used to explain how organic things grew, replicated and moved of their own accord, but typically vitalism and Aristotle's magical forms all came down to *some mysterious kind of push that makes life do lifelike things*.

In opposition to vitalist theories were *mechanist* explanations of nature. These tended to break living phenomena down into simpler, more fundamental theories of physics and chemistry. Vitalism was vague and qualitative, compared with the precise and mathematical descriptions of human anatomy based on biomechanics and biochemistry.

One persistent vitalist theory argued that many simple life forms could spontaneously emerge from the right blend of non-living materials. While it was obvious that most organisms required at least one parent, simpler life forms such as flies, lice or even mice were thought to pop into existence, if the right conditions existed. Some academics even provided recipes. The seventeenth-century Flemish chemist Jan Baptiste van Helmont advised that one could 'carve an indentation in a brick, fill it with crushed basil, and cover the brick with another so the indentation is sealed'. After exposing the two bricks to sunlight, you'll find the crushed basil will have turned into scorpions. Voila!

Want more fun? 'If a soiled shirt is placed in the opening of a vessel containing grains of wheat, the reaction of the leaven in the shirt with fumes from the wheat will, after approximately twenty-one days, transform the wheat into mice.'[14] So forget home brewing; give home mousing a go.

No doubt the apparently sudden appearance of plagues of flies, locusts or rodents could make spontaneous generation more believable than old-fashioned baby-making. Take some ordinary material, wait until it infuses with the vital essence, and presto! Maggots, mould or molluscs. Possibly inspired by Judges 14:8, which reads, 'And after a time he returned to take her, and he turned aside to see the carcass of the lion: and, behold, there was a swarm of bees and honey in the carcass of the lion,' bees were thought to form inside dead animals. Likewise, frogs formed out of mud, clams were made from sea foam, and salamanders were born in fire.

The theory had plenty of detractors. In fact, one of history's first recorded instances of a positive control in an experiment was developed to facilitate the critical questioning of this idea. The seventeenth-century Italian naturalist Francesco Redi doubted stories of maggots being generated by rotting meat alone, so he devised an experiment using half a dozen glass jars containing selections of fresh meat. Some of these morsels he covered with cheesecloth, and some he left exposed. A few days later, maggots writhed through the samples of uncovered meat, but were absent in the chunks of covered meat, demonstrating that the maggots had to be the result of a relatively large, intrusive visitor.

The proponents of 'vital essences' weren't put off by a single experiment, though. It simply pushed their front line back a touch further. If flies and mice didn't generate spontaneously, they'd say, then no doubt something simpler did, such as those recently discovered animalcules which had been spotted with those spiffy new microscopes.

Designing a controlled experiment to settle the question of whether microbes drifted in from the air or whether they could spontaneously grow out of sterilised broth demanded more than covering some jars with cheesecloth. Preventing a few flies from getting to meat was one thing; holding back floating clouds of invisible contaminants was a lot harder. Some kind of impermeable seal would have done the trick, but picky critics argued that the small critters needed a good supply of oxygen. An experiment like Redi's required an air supply that wouldn't also waft in any microbes.

Pasteur designed the perfect solution in the form of a glass vessel with a twisting, swan-like neck (or, *en français*, a *col de cygne*). The snaking entrance to the bottle allowed oxygen to enter the container while impeding dust motes or other drifting particles. Pasteur heated broth in his *col de cygne*, killing anything that already happened to be growing inside, and allowed it to sit for an extensive period. Nothing grew in the bottles of broth he left untouched. Meanwhile, in others he swished broth into the bottle's neck or snapped off the S-tube, introducing airborne contaminants which soon blossomed into colonies.

Just as Redi had shown that a new generation of flies needed an open channel through which an egg-gorged mother could sneakily enter, Pasteur demonstrated that microscopic life only grew when the contaminating parent particles had access. For his efforts he was awarded the prestigious Alhumbert Prize in 1862. Two years later, before an audience at the Sorbonne, Pasteur rubbished the very idea of the spontaneous generation of life: 'Never will the doctrine of spontaneous generation recover from the mortal blow of this simple experiment!'[Δ]

And it didn't. As a theory, ongoing spontaneous generation was dead in the water, its champions falling silent on the topic one by one. There was still the question of how life first arose on Earth, but few scientists now believed they could mix the right ingredients in

---

[Δ] The full quote is worth reading: 'And therefore, gentlemen, I could point to that liquid and say to you I have taken my drop from the immensity of creation, and I have taken it full of the elements appropriate to the development of inferior beings. And I wait, I watch, I question it! Begging it to recommence for me the beautiful spectacle of first creation. But it is dumb, dumb since these experiments were begun several years ago; it is dumb because I have kept it sheltered from the only thing man does not know how to produce, from the germs which float in the air, from life, for life is a germ and a germ is life. Never will the doctrine of spontaneous generation recover from the mortal blow of this simple experiment!' (Vallery-Radot, R. (1902) *The Life of Pasteur*. McClure, via archive.org.)

a bowl and discover microscopic organisms floating around a few days later. There was no longer a place for ghostly vital essences that seeded life.

As well as putting the final nail in the coffin of spontaneous generation, Pasteur's discovery that microscopic particles could spread, reproduce and befoul organic material cemented the contagion argument over miasmas. If yeast and yeast-like organisms could contaminate a body and affect the living chemistry of their surroundings as they multiplied, it wasn't far-fetched to believe that illness could result from similar infestations of 'germs'.

In fact, a whole 200 years earlier, chemists were pretty sure there was a relationship between illness and fermentation, which at that time simply described any form of digestion or rot. Take the English chemist and physicist Robert Boyle, for example. '[H]e that thoroughly understands the nature of ferments and fermentations,' Boyle wrote, 'shall probably be much better able than he that ignores them, to give a fair account of divers phenomena of several diseases (as well fevers and others).'[15]

Pasteur had set out to demonstrate how yeast didn't generate from vital forces and grape-guts, but came from the environment – namely, the skins of the grapes themselves. He went on to show that a silkworm disease called pébrine was also caused by a fungal microbe, establishing one of the first records of a disease caused by a microbial infection.

So the silkworms might have been happier, but the first human disease to be pinned to a microscopic pathogen didn't come until later. It was a team effort too – not that the Norwegian Gerhard Armauer Hansen and his German colleague, Albert Ludwig Sigesmund Neisser, would be overjoyed at sharing the honour of its discovery.

In his early days as a physician, Hansen had taken an interest in the biology of leprosy, which was considered by many at the time to be an inherited condition. Hansen thought otherwise, loudly proclaiming in argument with his teacher that it was in fact a contagious disease. The brash young doctor set out to prove it, studying the lymph nodes and other tissues of leprous patients under a microscope.

Inside what others had dismissed as degenerated fat cells termed 'brown bodies', Hansen thought he saw tiny rods. Sausage-shaped microbes weren't new to science – in fact, the term *bacteria* was from a Greek word for 'staff', and had already been used for the past twenty years to describe such tiny sticks. Hansen now declared these particular brown-body rods to be specific indicators of leprosy.

The challenge, the Norwegian knew all too well, was not just to show that these objects appeared in leprous tissues – he had to prove they caused the disease. That prescient German anatomist we met earlier, Friedrich Gustav Jakob Henle, had advised decades before that merely discovering potential *contagia animata* – or germs – in a sick patient wouldn't be good enough to prove they were the cause of the illness. These microscopic animals would have to be extracted and then grown outside the body as well. So when Hansen announced (correctly) to the world in 1873 that leprosy was caused by bacteria, he knew he was doing so without solid proof.

Meanwhile, a student of Henle's named Robert Koch had taken up his master's mantle, focusing his studies on the troublesome microbe suspected of being responsible for anthrax. He developed his teacher's advice into four rules, which can be paraphrased succinctly:[16]

1 The germ should be found in those suffering the symptoms, but not in otherwise healthy subjects.
2 The germ should be isolated from the host and grown separately.
3 The germ should cause the same symptoms when introduced to a healthy subject.
4 The germ should be found again in the new host, and then isolated, grown and shown to be the same germ.

Only when these conditions were satisfied could anybody confidently believe that a particular microbe caused a particular disease.

It's a nice idea on paper. In reality, these four 'postulates' are easier said than done. As good old Typhoid Mary showed us, pathogens can cause no symptoms in some hosts while causing illnesses in others, for example. Koch himself realised this in his later years, acknowledging that his rules had some kinks that needed ironing out. Isolating microbes in the late nineteenth century was also challenging – the leprosy bacterium Hansen discovered wasn't isolated until the 1950s, when the American microbiologist Charles Shepard finally managed to grow the finicky bug *Mycobacterium leprae* on the strangest of media: the footpads of mice.

It's debatable whether Koch's postulates helped or hindered the early search for microbial pathogens. If we stick to the four rules, we have to conclude that Hansen's discovery was little more than a lucky guess. In 1879 the Norwegian excitedly shared his tissue samples with Albert Neisser, a newly arrived student of Koch's, hoping the newcomer could apply his cutting-edge staining techniques to the slides and improve the resolution of his little rods, which might provide a few more clues to their behaviour.

Neisser did just this, and on return to his home town in Germany – with a handful of the newly stained slides as souvenirs – he published a paper on the discovery of the bacterium responsible for leprosy. All under his own name, too, with only the briefest nod to Hansen's work.

Of course, Hansen wasn't too happy about what he saw as a theft of his rightful claim to the discovery. History eventually rewarded the Norwegian for his dedication, renaming the illness Hansen's disease. And what dedication he showed! The physician went so far as infecting the eye of a patient – without consent – with tissue from another leprosy sufferer, just to prove it really was responsible for the disease. It should be noted that he was sued for this transgression.

Before we start feeling too bad for Neisser, he went on to describe the pathogen responsible for gonorrhoea, which was subsequently named after him – *Neisseria gonorrhoeae*. Never let it be said microbiology isn't glamorous.

By the turn of the twentieth century, the germ zoo was rapidly filling with new specimens, and their relationship with diseases was firmly being established. Advances in culturing and staining, along with further improvements in microscope technology, threw fuel on the bonfire of the revolution. The search for the contagions responsible for all manner of diseases was on, from syphilis to smallpox, cholera to rabies.

Success was scattered. An even smaller world of microbes seemed to lurk out of sight, beyond the reach of the most powerful microscopes. On successfully developing a vaccine for rabies in 1880, Pasteur began his search for the germ responsible for the disease. He eventually gave up, conceding that if an organism was responsible, it was too small to be visible beneath modern microscopes, and would be impossible to grow in a culture.

A ceramic filter designed by a fellow French microbiologist, Charles Chamberland, was subsequently developed, with Pasteur's help, to screen such sub-microscopic germs from liquids. This allowed researchers to show that, in many diseases, the residual fluid continued to be infectious while the filtered material wasn't. The term *contagium vivum fluidum* ('contagious living fluid') was initially used to describe the soluble disease agent.

An old Latin word for a poisonous fluid was then revived to describe this new category of contagion – the *virus*. Few suspected there were smaller particles than bacteria in the filtrate, so at first it was thought that infectious liquid toxins were dissolved in the solvent. A pair of German scientists, Friedrich Loeffler and Paul Frosch, argued that the 'soluble toxin' was indeed a particle in their analysis of foot and mouth disease in 1898.[17] While better filters and the invention of super-magnifying electron microscopes would show that viruses are a dazzling variety of organic particles, these odd occupants of a zone halfway between a non-living chemical and a simple cell have continued to surprise and fascinate pathologists.

In fact, modern-day pathology recognises a continuum of agents between the organic molecule and the macroscopic organism – from rogue zombie 'prion' proteins that bend other proteins into their twisted alter egos, to adventurous strands of RNA called viroids, which leap between plants, to chunky beasts of viruses that are massive enough to be seen with light microscopes, to a zoo of bacteria (and their cousins, the archaea), to protozoa such as the malaria-causing plasmodium, to branching fungi, and even to tiny worms.

The discovery of the germ allowed doctors to know their enemy. Gone were vital essences and putrefying smells. These foes weren't intangible forces of nature that corrupted an otherwise perfect state of health. They were solid entities, alive and therefore potentially destructible.

Within half a century germ theory went from nigh unbelievable to the only game in town. Late in the eighteenth century, an epidemic of heart attacks swept Asia as industrialised rice-processing reduced vitamin B1 (thiamine), leading to a condition called beri-beri. The scientific tools that could diagnose the real cause and remedy it did exist, but the medical establishment, primarily a Dutchman by the name of Christiaan Eijkman, was enchanted by the belief that beri-beri was caused by a microbial infection.[18] Only when Eijkman realised that his experimental chickens weren't being cured of a bacterial infection but were getting the nutrients they had lacked from a helpful handout of rice did it become clear that thiamine deficiency was behind the epidemic. It took nine years of experimentation for him to figure out that he'd been on the wrong track.

A far more gruesome infatuation with the pathological nature of germs took place at the New Jersey State Lunatic Asylum in 1907. Its medical director, Dr Henry Cotton, held the unconventional belief that the mental health of his patients was the result of infection. On the one hand, Cotton was rather progressive in his practice, avoiding putting his patients into constraints, reducing violence on the premises and introducing various novel therapies. Following the discovery of the bacterium responsible for syphilis, the good doctor theorised that germs were behind all forms of madness. By 1917 Cotton was ripping out teeth, gall bladders, uteruses and any other body parts he suspected might harbour the offending microbes. Patient or even family approval was rare, and his asylum's mortality rate was high. Sadly, Cotton's influence stuck around long after he passed away in 1933, with patient's teeth still being needlessly extracted as late as the 1960s.

Questions on the reach of the pathogen remain open for debate today, with researchers investigating whether various types of cancers, neurological conditions and auto-immune disorders may involve the presence (or absence) of certain viruses, protozoa, fungi and bacteria.

There's no doubt our health is entwined with the microbial world. Understanding that many of the most horrid diseases to have plagued humanity were caused by an occupation of microscopic organisms was a monumental victory. The concept has come to be called the *etiological principle*, and now forms the foundation of our treatment of infection.

The discovery of the pathogen grew to become synonymous with the war on disease itself. This invader was the enemy, not our own bodies. But for a truly decisive victory, we needed a powerful arsenal of weapons that would obliterate these tiny marauders for good, and ensure they could never again infiltrate our healthy bodies.

# INFLAMED

On the south side of the Thames, in the London borough of Southwark, is one of my favourite museums. You need to climb a narrow, winding staircase to get up to what used to be a section of St Thomas's Hospital and Church. In the mid-nineteenth century the hospital was moved across the river, and most of the wards, staffing quarters and operating theatres were demolished to make way for the Charing Cross railway line. With refurbishments made to house a post office downstairs in the late 1950s, one of the old theatres was rediscovered next to a garret used to dry herbs, part of what had once been the women's surgical ward.

Built in 1822, this operating theatre would have witnessed the brutality of what is rather brazenly called the heroic age of medicine. It was the era of the quick knife, of bloodletting and gruesome surgical interventions. Made bold by advances in anatomical knowledge, physicians progressed from old-fashioned barber/surgeons to become university-educated specialists.

Yet while medical practitioners were mapping the human body in ever greater detail, knowledge of infection remained medieval – if not worse. Take a tour of the museum and you'll be shown the old operating table left practically in situ after the hospital's relocation. A sign in Latin on the wall high above the table reads: *Miseratione non mercede* – 'For compassion, and not gain'.

Hospitals such as St Thomas's were no place for the sick and wounded. Only the most desperate would risk visiting one, perhaps if they lacked the funds to pay a doctor for a home visit. In front of a bloodthirsty crowd of paying customers – mostly students, although some from the city's elite sought an alternative to another night out at the opera – a poor, inebriated patient with rotting extremities, gaping abscesses, painful kidney stones, broken bones or other horrid afflictions would be held down beneath the surgeon's knife.

It's not hard to picture the stocky physician gripping the puffy arm of his patient, tourniquet already in place, a razor-sharp knife as long as a forearm held against the stretched flesh, the surgeon's arm poised to whip quickly around the limb the moment they're ready to operate. You can almost hear the patient's agonising cries reverberate against the skylight as the knife slips through skin, then fat, then muscle, exposing the pallid bone, which will take many seconds to saw through. Just maybe the drunk, panicked patient will be lucky enough to black out from the agony.

The surgeon slices, saws and stitches as quickly as he can, a tide of blood spilling into boxes of rags and sawdust under the table, stray trickles soaking into more sawdust packed between the floorboards. Arteries are tied off with catgut, the skin sewn shut with the same, and the entire wound dressed with torn-up sheets of old and probably dirty linen.

Speed was the key. One surgeon with a reputation for being quick with the knife was Robert Liston, who was reported to be able to lop off a leg in two minutes flat. His haste came at a fatal cost on at least one occasion, when he sliced not just through a patient's limb but also through the fingers of an assistant who was holding tight the muscle and skin. The patient didn't make it; meanwhile, gangrene set in to the unfortunate helper's digits, taking his life as well. By some accounts, a member of the audience dropped dead of a heart attack, making it a really bad day for poor old Bob.

When done for the day, the surgeon would hang up his gore-splattered leather apron (which wouldn't look out of place on a butcher), give his hands a quick wipe and head out on his afternoon rounds. The bloody garments were a mark of pride and surgical experience, so no thought was ever given to cleaning them.

Most surprising is not that any soul would risk such unhygienic torture in a place so foul as a hospital, but that anybody survived these operations at all. The records of the Glasgow Royal Infirmary in 1853 include an analysis of 284 surgeries over the previous decade: only one in six people died following amputation of the forearm; two in three died when the whole arm was removed; and patients only had a fifty-fifty chance of walking out on one leg.[19] The odds were only slightly better than even that you'd survive surgery at London Hospital in Whitechapel in the mid-nineteenth century.[20] That was actually pretty good; thanks to poor conditions, poor patients and poor luck, your chances of going home after surgery at many other institutions were well under 30 per cent. It might seem that the writing was on the blood-splattered wall.

Yet far from being accepted as a sign of poor health, ooze was seen as an indication that the body was functioning as expected. To Ignaz Semmelweis's fellow obstetricians, pus around the womb of a deceased mother was thought to be milk intended for the newborn. An absence of pus was seen as a bad sign, suggesting that death was imminent.

Since infection was inevitable in a world where surgeons didn't believe in so much as a cursory scrub, it's possible that an absence of purulence could have been the result of a dysfunctional immune response, spelling death. Put simply, where we'd view pus as a concern, in the early to mid-nineteenth century a milky build-up of dead white cells within malodorous crimson flesh was taken as a sign that the body was in good health.

Excluding religious-based practices, bathing and handwashing prior to the nineteenth century was a luxury performed in the name of comfort or aesthetics – a sign of class, rather than an attempt to avoid illness. Baths in Roman times, as in many pre-medieval Western cultures, were social affairs, to the point that they were often associated with debauchery. Partly for this reason, Church-based doctrines taught that bathing risked sin. Cleanliness wasn't close to godliness at all; by the fourteenth century, dirt was seen as a sign of humility and health. In fact, at times it wasn't uncommon for immersive bathing to be avoided altogether, for fear of disease, as it was generally thought that water opened the pores and weakened the body, making it susceptible to sickness.

Even the famous British nurse Florence Nightingale – pioneer of hygienic conditions for bedridden soldiers during the Crimean War – believed sickness was the result of stuffy, grotty conditions fouling the air, not infectious particles. She advocated that nurses wash their hands, but not because she believed in contagions: it was because the sweat and grime itself was toxic. '[I]f the sick are to remain unwashed, or their clothing to remain on them after being saturated with perspiration or other excretion,' she wrote, '[a nurse] is interfering with the process of health just as effectively as if she were to give the patient a dose of slow poison by the mouth.'[21] Nightingale did not believe there was such a thing as specific diseases. Sickness was physiological rather than ontological; it was what the body did, not a 'thing' that happened to it.

It wasn't unusual in previous centuries for physicians to treat all manner of symptoms as personal responses to a disruption in health, and not as discrete diseases one could catch. Nightingale herself asked: 'Is it not living in a continual mistake to look upon diseases, as we do now, as separate things, which must exist, like cats and dogs?' Taking the battle against germs from Pasteur's flasks to the hospital bed would continue to face stubborn opposition from a conservative medical tradition for many years.

Fortunately, the fight was one an Essex lad by the name of Joseph Lister found compelling. As a young surgeon, Lister was fascinated by inflammation. By the latter half of the nineteenth century the puffy redness of traumatised tissue was considered a disease in its own right – a further indication of the fuzziness of the borders between illness, symptom and disease. While working as a dresser to the eminent surgeon John Eric Erichsen, Lister was taught that wounds emitted their own miasma, promoting inflammation and poisoning not just the surrounding flesh of that patient, but of all others in the ward. Erichsen advised a balance of patients in a room, fearing that too many wounds would create a miasma that would ultimately spread out of control.

Lister wasn't convinced by his teacher's thoughts on corrupting gases. He noticed that cleaning and debriding a wound (clearing away dead and dying tissue) as he changed its dressing reduced inflammation and improved its chance of healing, which suggested that physical material could be playing a more direct role. It was only later, on reading Pasteur's paper *Recherches sur la putréfaction*, that Lister realised inflammation was caused by microbes on the skin entering the moist environment of the wound, not unlike yeast on the skin of grapes. Semmelweis's work on cadaverous particles was familiar to Lister, as were the American physician Oliver Wendell Holmes's theories on contamination, but it took Pasteur's experiments and Lister's own experience to bring the pieces together.

Inspired by a newspaper report on the use of creosote in reducing the stench of sewage, Lister experimented with carbolic acid to disinfect wounds. His success was as clear as day, and in 1867 Lister enthusiastically took his findings to the medical establishment. He described in *The Lancet* that, prior to introducing his carbolic acid treatment, 'the two large wards in which most of my cases of accident and of operation are treated were amongst the unhealthiest in the whole of surgical division at the Glasgow Royal Infirmary'. Since his antiseptic treatment, the wards 'have completely changed their character; so that during the last nine months not a single instance of pyaemia, hospital gangrene or erysipelas has occurred in them'.[22]

Lister created a carbolic acid spray that could be used to disinfect an operating table. This was met by surgeons with a mix of scepticism and jubilation, yet the survival rates spoke for themselves, and the next generation accepted the merits of antisepsis in surgery without question. Sterility in the operating theatre gradually encouraged the taking of greater surgical risks, opening the way for dangerous new procedures that had previously been off-limits, such as digging around inside the chest or abdomen.

Globally, the change in health culture was anything but instantaneous. Germany adopted the new practices relatively swiftly, while Great Britain dragged its heels. By the turn of the twentieth century, however, hospitals across the world had ditched rancid aprons for white frocks, recycled bedsheets for pristine bandages, and infrequent mopping for bleach and detergent. The new priority was to prevent germs getting near the patient in the first place.

But what of those microorganisms that made it through the hostile landscape of antiseptics? Harsh chemicals such as acids and bleaches do a fine job murdering microbes by the billions, but once an infection has established itself deep inside the body, hitting it with anything corrosive risks obliterating healthy human tissue as well. What was needed was a magic bullet that would shoot germs without harming their host.

A German physician named Paul Ehrlich believed such a chemical was possible. His expertise in developing stains that were specific to particular tissue types and microbial organisms hinted at chemicals that might affect some cells and not others. Quinine was understood to treat malaria without damaging a patient's own cells; Ehrlich just needed to find chemicals that could do the same thing with other pathogens.

His ambitious goal was to develop a drug that could kill all diseases – what he called in German a *magische Kugel*, or 'magic bullet'. His ground-breaking work led to the discovery in 1909 of a substance he called compound 606, or arsphenamine, which killed the spirochaete bacterium responsible for syphilis with relatively minimal impact on the host. The drug was marketed the following year as Salvarsan, and proved far more effective than the mercury-salt-based drugs that had previously been used against the disease.

As revolutionary as Salvarsan was, not everybody celebrated its development. To the more socially conservative members of the public, the availability of such a drug was thought to encourage promiscuity. As we've seen time and again over the years – think of the contraceptive pill, mifepristone (also known as the abortion pill RU-486), HIV pre-exposure prophylaxis medications, and even the humble condom – any attempt to medically reduce the unwanted consequences of sexual intimacy is thought by some to encourage deviancy. As a result of developing Salvarsan, Ehrlich and his Japanese colleague, Sahachiro Hata, were vilified and accused of criminal conduct, with their research put under the public microscope.

Despite its success, in the end Salvarsan wasn't the *magische Kugel* Ehrlich sought. The discovery of the ground-breaking antibiotic that would change the world was still several decades away.

The Scottish pharmacologist Alexander Fleming's chance observation of the mould *Penicillium notatum* growing in a Petri dish is a classic tale of scientific discovery. Like many stories of scientific innovation, Fleming's famous discovery amazes us with its sheer fortune. Invention of the drug that has saved literally tens of millions of lives (if not hundreds of millions) began not with 'Eureka!' but with 'WTF?'

Also like much in the history of science, however, the true story of penicillin's invention is a little more complicated than our high-school textbooks would have us imagine …

Among the world's great variety of traditional remedies for wounds and infections are poultices and dressings based on materials we might consider decaying, rotten or spoiled, such as mouldy bread or off milk. In Australia some Indigenous cultures made use of mould taken from eucalyptus trees. A seventeenth-century English apothecary to the king named John Parkington recommended treating wounds with mould. Even the Jewish Talmud suggests mouldy corn soaked in date wine – or *kutach bayli* – as a cure for infections. Though it's questionable whether these remedies were effective (let alone effective due to the antimicrobial properties of their resident microbes), practices such as these hint at a history of observation of the silent chemical warfare between competing species of microorganism.

Fleming wasn't even the first scientist to note the particular antibacterial talent of the *Penicillium* fungus. As nineteenth-century microbiologists tried, often in vain, to isolate various microbes using all manner of broths and gels, they regularly discovered that recipes would fail if they were contaminated. For instance, Pasteur knew that urine was useful for growing the anthrax bacillus, unless it was already home to other species of oxygen-loving bacteria.

The renowned British naturalist Thomas H. Huxley – a.k.a. Darwin's Bulldog – found in 1875 that bacteria didn't fare well around the fungus *Penicillium glaucum*. Although he shared this discovery with the physicist John Tyndall – a friend of Lister's and a penpal of Pasteur's – the observation went no further. Tyndall was better known for his studies on light passing through the atmosphere, yet the scientist's efforts to create containers of particle-free air led him to experiment on airborne bacteria. Both Tyndall and Huxley were therefore in a good position to spread the word about the antibacterial properties of *Penicillium*, well over half a century before Fleming's discovery. They didn't.

Their failure to seize a golden opportunity to discover a germ-fighting super weapon may have been a case of missing the wood for the trees. In the 1870s, Tyndall was a self-declared marshal against the spontaneous generation cause. His work on thermodynamics and the conservation of energy put him in opposition to the wishy-washy world of vitalism. Like Tyndall, Huxley was a mechanist through and through, promoting Darwin's theory of natural selection as an explanation for the world's biodiversity. Germ theory was, for Tyndall, ammunition in the great battle between those who believed biology had no need of unique physics and those who felt life required its own force. It was not for curing diseases.

Huxley and Tyndall were so preoccupied with putting the boot into proponents of vitalism and spontaneous generation that their observations of *Penicillium* fell by the wayside. Whether it would have sparked an early inquiry into a broad-spectrum antibiotic is anybody's guess. Yet, to paraphrase British historian James Friday, them's the breaks in science: 'If scientific controversy accelerates research and clarifies ideas, it may also leave a number of loose ends, any one of which, in more settled surroundings, could have been the subject of separate inquiry.'[23] Of course, it wasn't until 1939 that Fleming noticed the effect of this genus of mould on nearby colonies of bacteria, and speculated that it was caused by an unknown chemical he dubbed penicillin. It took a team of chemists headed by Australian Howard Florey to isolate the compound and test its effects.

In 1941 Florey and the German-born Ernst Chain administered a dose of penicillin to an unfortunate patient with a severe infection of the face and scalp. Although he showed signs of rapid recovery, the chemists hadn't extracted enough and he passed away. While this was sad for the patient, the scientists now had clear proof of penicillin's magic.

Penicillin works by interrupting the pathway a bacterium uses to build a cell wall, forcing its gooey insides to continue to outgrow its now frozen shell. Given the variety of microbes that use the same wall-building process, and the fact that we animals don't build cell walls, it's a rather effective weapon against bacteria. Modifications to the penicillin molecule have broadened the range of susceptible bacteria even further. Across Europe, penicillin-based antibiotics account for between 32 per cent and 67 per cent of all antibiotics prescribed, thanks to it being the most wide-sweeping 'magic bullet' in our antibiotic arsenal.[24]

Antisepsis and antibiotics had landed a solid one-two punch on microbial disease – it was a medical revolution. Now, the etiological principle means physicians can diagnose pathogens and identify likely treatments, while half a century of innovation in chemistry has produced a wide arsenal of antibiotics that take care of the vast majority of debilitating bacteria.

The knockout blow for many pathogens, however, can ultimately be traced back several centuries to a practice in the Near East. As with many technologies and scientific theories, though, following a modern medical practice down its path of inheritance doesn't necessarily show you a clear historical genesis.

In 1721, while living in what is today Istanbul, the English aristocrat Mary Wortley Montagu enlisted the help of the English embassy physician Charles Maitland to copy a practice she'd seen among the Ottoman civilians, which she described as 'engrafting'. It involved finding a smallpox sufferer who seemed to have mild symptoms, and inserting fluid from their blister into a healthy individual. Since it was considered rare for a person to catch smallpox twice in their lifetime, catching a mild case of the disease was much better than catching a bad one. Lady Montagu had her son inoculated in this way in Istanbul; several years later in England, in the face of a smallpox outbreak, she asked the same physician to inoculate her daughter as well.

The practice of what came to be called *variolation* spread through England. It increased immunity across a community, yet wasn't without risk to individuals. Mild symptoms could be caused by weakened strains of the smallpox virus, although subtle differences in immune systems could also account for the disease's impact.

Variolation has an extended history. The physician Edward Jenner is credited with tweaking the practice to make it slightly safer. Observing that the women who contracted cowpox tended to avoid the ravages of smallpox epidemics, in 1796 Jenner reportedly scraped fluid from a cowpox blister on the hand of a milkmaid and jabbed it into the arms of the eight-year-old son of his gardener.[Δ] This practice of *vaccination* – named for the Latin word for 'cow' – reduced the risk of infecting a patient with a virulent disease, allowing us to confer immunity to specific pathogens at a personal risk far lower than the disease itself.

Vaccination contributed significantly to the extinction of the smallpox virus. Never again will the world see populations decimated by this horrid illness. Poliovirus and measles stand on the edge of eradication, gone but for occasional outbreaks among parts of the population where vaccination rates are low. Research is making headway on developing vaccines for malaria, HIV and many other microbial diseases that continue to threaten human lives.

---

Δ   A recent analysis of the few remaining vaccines Jenner had prepared revealed that none contained any traces of the cowpox virus, but rather a virus that is more likely to have come from horses. It's possible he was experimenting with a variety of poxes, of course. But perhaps we should be referring to vaccinations as equinations? (See Damaso, C.R. (2017) Revisiting Jenner's Mysteries: The Role of the Beaugency Lymph in the Evolutionary Path of Ancient Smallpox Vaccines. *The Lancet Infectious Diseases*, vol. 18 (2), pp. e55–e63.)

A combination of sanitation, quarantine, antibiotics and antisepsis, along with a plethora of antiviral and anti-parasitic pharmaceuticals, has helped humanity conquer the germ. Reflecting on the countless lives lost to infection and plague through history – diseases such as typhus often saw more soldiers to their graves than enemy weapons, while populations have been annihilated by bacteria such as *Yersinia pestis* and viruses such as influenza – it's hard not to see the vanquishing of infectious disease as the greatest triumph of modern medicine.

The devastating impact of the germ, followed by its monumental defeat, has defined disease in general as not just contagion, but as external, mechanical and vulnerable. Diseases have become Florence Nightingale's 'cats and dogs'. All of this has truly changed the world for the better. We owe a great debt to Pasteur and Snow, Lister and Semmelweis, Neisser and Hansen, Montagu and Jenner, Ehrlich and Hata, Nightingale and Fleming, and a gazillion other minds of various degrees of notoriety. Waging war against such a diverse enemy is worthy of celebration.

But military metaphors can only ever be so appropriate. Microbes as a whole don't exactly pose a unified threat. If anything – even keeping in mind the suffering many bacteria, protozoa, fungi and viruses have caused – our relationship with the unseen world of the germ is more neutral, or even allied, than oppositional.

# INHABITED

Ancient Earth was a bug-eat-bug kind of world. The first organic blobs we could possibly describe as alive were little more than envelopes of fat sheltering short threads of nucleic acid – what the physicist Freeman Dyson colourfully called 'garbage bags'.[25] It's unlikely that at first there were clearly defined species – at least, not as we might think of them today. Just countless stunted sentences of RNA codes bumping and grinding, decaying and building, blindly competing without aim or direction in a rich soup of carbon-based compounds.

Some specific RNA sentences would have been better than others, therefore giving their garbage bag some kind of advantage, protecting them for just a few more precious minutes or copying themselves just a little bit more vigorously, until over time – a long time – these fat bubbles became consistently complex life forms we might compare with modern-day organisms we call *prokaryotes*. Over several hundred million years or so, distinct lines of inheritance emerged, isolating families of prokaryote into two obvious bunches of bugs: *bacteria* and *archaea*.

Superficially, it's easy to confuse these two groups, what with their absence of a nucleus and relatively basic internal organisation. In fact, prior to 1977, taxonomists grouped all members of these two massive tribes under the single banner of 'bacteria'. Yet biochemically speaking, archaea and bacteria are like the Androids and iPhones of the primeval biosphere.

There is no place on our world's surface that these two domains of life haven't inhabited, from the frozen poles to deep underground where light can't reach. Archaea blossom in acidic ponds. There are microbes living in the boiling water of deep ocean vents. No place is too hot, too cold, too toxic, too acidic. What's more, both archaea and bacteria literally shape our world. Our atmosphere is oxygenated thanks to their ability to rearrange carbon dioxide and water to grow. Their secretions bind rocks, their enzymes weather mountains, their bodies precipitate rain. For all their simplicity, these single-celled organisms have ruled Earth for most of the planet's existence, and will continue to do so long into the future.

Like modern bacteria and archaea, early life forms would have been ravenous scavengers, evolving ways to pocket stray scraps of chemistry that advantaged their ability to survive and replicate in a dangerous, highly competitive environment. Typically, this trash harvest absorbed short, random sequences of stray nucleic acid floating by, pasting it into their personal library on the off chance it would be useful.

In a more dramatic example, some organisms have engulfed entire cells in ways that allow the prey to stay alive rather than succumb to digestion. Ultimately, we are the result of one such dastardly act of cytological kidnapping. All plants, animals, fungi and protists are made up of compartmentalised cells called *eukaryotes*; vestigial forms of this ancient bacterium are seen inside each of our cells in the form of the power-transforming units called *mitochondria*. Another example of prokaryotic slavery persists in plants as green-pigmented chloroplasts, which were once free-living, algae-like cyanobacteria. These two marriages between otherwise simple cells paved the way for life to evolve into cooperative mega-colonies, giving cells an energy boost that allowed them to specialise in jobs that helped their cellular city survive. Life blossomed in union as much as in competition. In some ways, we are the product of a momentous infection that's persisted for some 2 billion years.

Not only do we continue to rely on the ancient contract between mitochondria and their host, our body's entire surface (inside and out) is now home to between 10 trillion and 100 trillion microbial cells, consisting of as many as 10,000 species, many of which are vital to our existence. This myriad of microscopic denizens, known as a *microbiome*, forms an ecosystem so complex that we're only beginning to tease out the details of its interactions and understand the role they play in our health and wellbeing.[26]

Our tiniest citizens do more than just passively occupy space that deters pathogens from setting up shop. They actively manage the development of our immune system, they turn our food into nutrients (which we're otherwise incapable of producing ourselves), they affect how we smell, and they possibly play a vital role in the communication between our guts and our nervous system. We do more than use our microbiome; we are so dependent on it that we might as well consider the microorganisms inhabiting our body to be an organ as complex and important as our skin, liver or pancreas. Our bodies have evolved to do more than tolerate microbes – our physiology actively encourages them. Even the once-belittled human appendix is now thought to act as a kind of intestinal bomb shelter to protect a selection of microbes, should an apocalyptic catastrophe strike our gut.[27]

Then there are the shadowy squatters among our very genes. Our chromosomes hold a graveyard of skeletal viruses, their now defunct codes buried among our own DNA sequences. Among them are also the viral zombies – sleepers that hide in our cells, ready to awaken and continue replicating using our biochemical machinery. *Herpesviridae* is an example of a latent episomal virus, better known as the family containing chickenpox, cold sores and genital herpes. This sneaky pathogen leaves its genetic material in the cytoplasm of our nerve cells, ready to awaken when conditions are in its favour and pump out batches of virus particles.

Far more benign, however, are the multitudes of viruses that cling to the thin layer of mucus covering our delicate inner surfaces, such as our lungs and digestive tract. These microbes aren't interested in invading our body – it's the bacteria they seek. As strange as it seems, even microscopic organisms can get infections.

Viruses called *bacteriophages* look a little like tiny lunar landers. They use their crooked legs to grab hold of bacteria and jam a thin tube through their wall and membranes. Like a mosquito in reverse, the phage spits its genetic code into the cell, commandeering the host to construct a new generation of phage bodies. The virus's presence contributes an additional layer of protection between our community of microbes and the sterile inner sanctuary of our blood and organs.

While a mere hundred or so species of bacteria are responsible for disease, we're slowly discovering multiple thousands of others that play vital roles in how our bodies function. And if we look at what microbes do in other species, we suddenly realise how casual our human affair is with the microscopic world.

Countless species of multicellular organisms are so dependent on microbes to break down food, release nutrients, fight invaders, guide development and control reproduction that they could not survive without them. Termites would perish without the cellulose-breaking protists *Trichonympha* and *Mixotricha*, which break apart the woody carbohydrates in their food. Corals bleach and die if their algae are expelled, leaving the tiny polyps with less food and exposed to other invading pathogens. The wasp *Asobara* can't reproduce at all without the help of a bacterium named *Wolbachia*.

This isn't to say germs are our friends – at least, no more than we can say viruses, bacteria and protists are predominantly our enemies. Like any species, microbes perpetuate their genes by taking opportunities to secure nutrients and safety within a given environment. Such opportunities are all about give and take, about forming tenuous alliances with other species. Sometimes the benefits are biased, sometimes they're reasonable. Many are stable, long-term agreements. A few are fleeting and selfish.

Pathology is less about asking *which* microbes are dangerous than about asking *when* a microbe might be dangerous. To illustrate the point, pathologists refer to a conceptual model known as the disease triangle, which illustrates a stable relationship between the host, their environment and the microbes within it.

Say a person loses their immunity or suffers some kind of injury. This upsets the balance of the triangle, allowing a microbe to access new grounds and cause problems. Anybody who has suffered from a urinary tract infection knows the impact of *E. coli* bacteria moving from their usual place in the gut to an otherwise sterile environment. Those having cancer treatment or who have had organ transplants must take care that their immunosuppressive medication doesn't let down their body's security screens and allow their microbiome to move beyond its bounds. If the host is exposed to new disease-carrying agents, a microbe could exploit the new territory. Bacteria that is quite comfortable in the digestive systems of poultry or fish, such as *Salmonella*, is devastating if it finds its way into our gut, for example. Lastly, if a microorganism evolves a dangerous new talent, such as a new kind of resistance or a toxic secretion, it can provoke symptoms where before it did not.

Bacteria can swap chunks of genetic material described as *pathogenicity islands*.[28] These encode novel ways of exploiting their environment, typically with negative consequences for the host. One example is the corkscrew-shaped microbe *Helicobacter pylori*. In 1982 Australian physician Barry Marshall famously took one for the team by downing a dose of the bacterium after both he and his colleague, Robin Warren, noticed an association between the microbe and peptic ulcers. Previously, stomach ulcers were considered to be a consequence of stress. Marshall's dramatic demonstration and follow-up research earned him both public attention and a Nobel Prize in Physiology for proving a link between the pathogen and ulceration of the stomach wall.[29]

Nearly half the world's population carries the *Helicobacter pylori* bacterium, although barely 15 per cent of these individuals develop ulcers. Yet for some reason, in a relatively small number of people, the *Helicobacter* bacteria living beneath their mucus produces a spot of inflammation, opening the way for a cascade of complications precipitated by corrosive stomach acid and the pathogen's own protein-breaking enzymes. Exactly why this occurs in some people and not others isn't clear. A clue might lie in the fact that one strain of *Helicobacter* with a particular pathogenicity island seems to be overrepresented among sufferers of peptic ulcers. The genes in this island code for a disruptive protein called CagA, which the bacterium injects into the cells lining the stomach, ripping apart a bunch of important molecules.

Pathogens aren't necessarily agents of disease. It just isn't that simple. A few, of course, can't help but be nasty. Many shift back and forth on the symbiotic spectrum between mutualism and parasitism as their environment and genes dictate. Many provide benefits as often as they do complications.

Although associated with ulcers and stomach cancer, *Helicobacter pylori* may help some populations by reducing the incidence of gastric[30] and oesophageal cancer,[31] reflux[32] and perhaps even asthma.[33] It's hard to know whether to cheer the reduction of ulcers or cry over the potential increase in reflux and asthma. Research indicates having resident *Helicobacter* bacteria in your gut neither reduces nor extends your lifespan by an appreciable amount anyway.[34]

Antibiotics and antisepsis have changed the world in immeasurable ways. For the most part, this has inarguably been for the better. Most people on this planet know of somebody who would not be alive if not for sterile hospitals or treatments that destroyed a life-threatening bacterial infection.

Drawing an equivalence between germs and disease has carried with it an unintended but nonetheless serious consequence, however. Indiscriminate use of disinfectants and antibiotics in an effort to obliterate 'germs' has produced a great deal of collateral damage – damage so significant it could unravel the very benefits scientists have worked so hard to produce.

# CLEAN

It's really a miracle I'm alive.

When I was an infant, my father brought home a small puppy, who was given the name Patch after the brown spot surrounding one eye. Tragically Patch wasn't long for this world, passing away from a mysterious illness but a few short months later.

Then there was the budgie I had when I was ten, creatively named Tweetie, who died in the middle of a violent thunderstorm not long after moving in, possibly from shock.

Susie the guinea pig belonged to my sister. If anybody knows what happened to Susie, let us know. Our money is on aliens.

Oh, and then there were the fish. All of them. Goldfish of all varieties. None surviving more than six months, until finally, defeated, we stored the tank in the back shed.

We gave up keeping pets at that point, for fear of PETA banging on our door. We were a cursed family as far as animal companions went. Just how my mother kept three children alive and breathing is the real mystery, to be honest. But had we risked keeping a few kittens around, science says we'd have been rewarded for it – if I had a variant of the gene 17q21, that is.

I suffered bronchitis as a kid that made me bark up my lungs every winter. Nothing serious, but it did require courses of muscle relaxants and steroids to ease the wheeze. If we hadn't been a poster family for pet cemeteries, letting a cat hang around my crib when I was born might have mitigated the risk of allergy conditions imposed by gene 17q21.[35] Another study suggested cats help lift levels of *Ruminococcus* and *Oscillospira* bacteria around the home, which could play a role in reducing the risk of allergies and obesity.[36]

What might my annual hit of hay fever be like if I'd been allowed a kitty as a kid?

A handful of cherry-picked studies aren't exactly solid proof that my mother is to blame for my September sneezes, but on balance there's sufficient reason to believe a world sanitised of pet dander, microscopic bugs and general filth is not as healthy as we might like to believe.

Visiting an Amish community is about as close as today's American citizen can get to stepping back into the mid-nineteenth century in the rural United States. Descended from a Swiss Anabaptist tradition, baptised members of the community strictly adhere to the 'Ordnung', the rules of their simple dairy-farming lifestyle. This includes limited use of modern technology, the wearing of plain clothing and a firm commitment to manual labour.

Allergist Mark Holbreich noticed that the Amish he served in Indiana seemed to have a lower incidence of immunity-related conditions. Working with a team of researchers based in Europe, Holbreich found the incidence of asthma among Swiss children raised on farms similar to those of the Amish to be just under 7 per cent. Among the Indiana Amish themselves, the incidence was about 5 per cent. Compare that with children raised in Swiss cities, who suffered around one and a half times as much asthma: 11.2 per cent. Hay fever was even more drastic – it was around 12 per cent for the urban Swiss kids, compared with just 3 per cent for their nearby farm cousins. And those Indiana Amish? Again, much lower, at just over half of 1 per cent. Allergy-based eczema, meanwhile, came in at around 12 per cent for the city-dwelling Swiss children, nearly 8 per cent of farm-based Swiss kids, and a tiny 1.3 per cent of Amish offspring.[37]

The message is simple: either something about city living is giving us allergies, or something about country living is helping us fight them.

The Swiss/Amish case study, along with the research on the role pets play in our developing immune systems, are just a few examples in a mounting pile of evidence supporting what has come to be called the *hygiene hypothesis*. In short, it suggests that our hypervigilance against germs may come at a cost to our health.

With antibacterial sprays that promise the eradication of 99 per cent of germs, our liberal use of hand sanitiser, our fear of letting infants shove a fistful of nature into their mouths, our urban distance from soil microbes, pollen and animal dander, our avoidance of close contact with other people, our chlorination of our water and our obsession with hitting every sniffle with an antibiotic 'just in case', many of us have radically altered the microbial ecosystems that surround us. This seems to have confused the hell out of our immune system.

While a typical textbook analogy of the immune system describes a defending army keeping watch on the castle wall, a more accurate one portrays customs officers at a border, checking passports and determining who stays and who goes. During development, our immune system is geared to pick up useful microorganisms, while keeping out those that could cause problems. Our tiny border police are trained to develop their identikits, fine-tuning appropriate red-alert inflammatory responses, fevers and mucus production. The hygiene hypothesis contends that without this training our immune responses become disproportionate, even for relatively innocuous invaders. This is thought to have led to an increase in conditions such as asthma, inflammatory bowel diseases and a variety of allergic reactions.

Our obsession with cleanliness has evolved into something more than just a keen awareness of potential pathogens. It has become a means for us to divide people, signalling differences in cultural categories of socioeconomics, education and morality.

As far back as the 1930s, advertising warned women not to let their husbands detect their body odours. Fragrant vaginas, armpits or feet stopped being considered healthy and became indicators of poor self-care, prompting daily (if not twice daily) showers, liberal applications of perfumes and deodorisers, and frequent clothing changes.

Homemakers are today encouraged to disinfect the family home to hospital standards. Animations of nameless bacteria multiplying beneath a microscope represent threats to health and happiness. More often than not, mothers are the main target: the marketing culture of the twentieth century has sought to make every woman a potential Florence Nightingale, using cleanliness as an indicator of morality. Antibacterial soaps and wipes are now a standard in just about every kitchen cabinet. Where once asepsis was a luxury, it is now a basic standard.

Although more than a century has passed since Lister developed his carbolic acid spray, antibacterial soaps are a fairly recent addition to the average domicile. The antiseptic hexachlorophene was used as an ingredient in germ-killing soaps from the 1940s. Unfortunately, in the 1970s it was associated with neurological abnormalities in infants,[38] prompting authorities to remove it from public use. It wasn't until 1984 that a patent for adding the bactericidal chemical triclosan was filed in the United States, replacing hexachlorophene as a popular soap additive. Since then, three-quarters of antibacterial liquid handwashing detergents and a third of soap bars have contained this hospital-grade disinfectant, promising freedom from illness-causing germs. An outbreak of avian influenza in 2009 prompted a spike in the use of antibacterial handwashing detergents, and it has remained at an all-time high ever since.

As of 2016, however, triclosan in the United States has seen its day. Along with eighteen other chemicals, the American Food and Drug Administration (FDA) has banned triclosan from being added to publicly marketed soaps.[39] According to the FDA, adding triclosan to soaps does little to rid people of potential pathogens. For one thing, most people don't wash their hands with the same rigour as surgeons in a hospital, or with the same amount of soap. For another, triclosan is only deadly to bacteria when left in contact with the skin for at least two minutes. Who has time for that on the way out from doing their morning number twos? In addition to being ineffective at banishing bad bugs from your hands, studies suggest it can affect vital organs in animals.[40] And when washed into local environments, triclosan can mess with the microbiology of other ecosystems.[41] In short, if it does nothing and poses even a small risk of unwanted consequences, why use it at all?

Far more alarming is the overuse of antibiotics, which leads to the creation of so-called superbugs. The acronyms MRSA, VRE and CRE, for example, describe common bacteria that have evolved a few chemical tricks to reduce or inhibit the effects of powerful antibiotics such as vancomycin or methicillin. Given that evolution knocks out the weak and favours the gifted, this was bound to happen sooner or later. Yet using antibiotics on crowded livestock as a prophylactic against *possible* infections, coupled with its often unnecessary use for undiagnosed viral conditions, has hastened the selection for resistance factors in these microbes.

Bacteria have a talent for exchanging genes like trading cards at a swap meet, so a single resistance can jump not just between individuals but between species and even genera. Between 2000 and 2010, the global use of antibiotics rose 30 per cent, giving a lot of individual superbugs the chance to shine.[42] In 2013 some 23,000 deaths in the United States and 25,000 deaths in Europe could be attributed to resistant pathogenic bacteria.[43] The race is on to find new antibiotics, as well as new classes of drugs that might slow the arms race between us and pathogens. Meanwhile, we face the bleak prospect of returning to a time when sepsis was a life-threatening condition, when pneumonia or even a simple cut could turn deadly.

It'd be stupid to romanticise the days of yore and eschew antibiotics and alcohol wipes. Nobody wants to go back to unwashed surgeons collecting blood patterns on their leather aprons, or to seasonal outbreaks of gastro thanks to faeces-tainted tap water. Even the agrarian ways of the Amish have little to teach the vast majority of us. While some of us might be able to afford a change of scene and a simpler lifestyle milking cows or tending crops, civilisation is increasingly urbanised; even owning a few chickens is a luxury for most of us city slickers.

Our world changed when humans aggregated into tight-knit clusters. The disease triangle suffered its biggest shake-up as populations of humans built permanent structures, and then defecated, raised livestock and sought nutrition all in the same space. Health risks of one type – such as the dangers of predation, exposure and malnutrition – were exchanged for others: plague, pox and pestilence.

Ironically, we wouldn't have science without the benefits of civilisation either. The ability to identify infectious diseases and then develop tools to defeat them relied on people having the resources to sit, study and communicate ideas. Such resources demand wealthy patronage, either from invested benefactors, charity or a market, each of which would have its unique values, and prioritise which pathogens to discover, treat and vaccinate against. One only need look at the influence of religious authorities on AIDS education in developing countries, downplaying or contradicting the evidence that condom use limits the spread of HIV.

Following the 2014 outbreak of Ebola, an epidemic that claimed around 5000 lives, the then director-general of the World Health Organization, Margaret Chan, cited profit as the primary reason vaccines were slow in being developed. 'A profit-driven industry does not invest in products that cannot pay,' she said. 'WHO has been trying to make this visible for ages now. Now people can see for themselves.'[44]

That might seem like a cold view of modern medicine. If 5000 people were to suffer horrible deaths in London, Sydney or New York, a vaccine might be a little more forthcoming – if not for want of profit, then at least out of shock at such a grotesque epidemic right on our doorstep.

The political relationship between scientific authority and the treatment of disease has also led to the rise of the vaccination sceptic. From conscientious questioners to committed opponents of vaccination, a robust and diverse counterculture has encouraged hesitation, if not outright abstinence, in providing children with the regime of inoculations deemed necessary by health experts. Objectors' reasons vary from concerns over harsh-sounding chemicals to a selfishness in avoiding risks while retaining benefits, to a belief in the virtues of immunity gained from 'natural' infection. Common to most, though, is a mistrust in the authorities who promote the importance of vaccination for the greater good.

Our appreciation of what constitutes disease, along with how we should treat it, is filtered by numerous voices, each of which has different agendas, priorities and definitions of what constitutes clean, good and normal. Recognising the impact these social values have on our definition of health and disease doesn't mean abandoning what works. It means understanding the context of our successes, and questioning our assumptions about what is normal in light of past and present cultures.

Contagion theory presumed that we exist in a prime state of health, until a toxic outsider infects our body and causes suffering by its presence. It's the role of the physician to identify the agent and prescribe a course of action that will remove it, returning the patient to the expected standard of normality. There is no overstating how powerful this system has been in giving us the ability to predict the unpredictable and control the uncontrollable. Germ theory, and with it the predominance of the mechanical over the metaphysical, is one of science's greatest triumphs.

There have been costs, though. The model has left a legacy that frames disease as objective, ontological, material, discrete and corruptive. It's also raised questions over whether addiction, obesity, fatigue, a lack of attention, and dysphoria over sexuality and identity qualify as true illnesses or are really deficits in our personalities.

# III
# INSANE

# ADDICTED

'Only one drug is so addictive, nine out of ten laboratory rats will use it ... and use it ... and use it ... until dead.' Cue sinister shadows surrounding a damp-furred rodent, which staggers about its cage in a confused montage, sniffing and twitching, until finally its poor, drug-poisoned body falls still. 'It's called cocaine,' the television advertisement's menacing voiceover reveals. 'And it can do the same thing to you.'

The 1980s campaign run by Partnership for a Drug-Free America – now called Partnership for Drug-Free Kids – found the perfect mascot in the poor old lab rat. Endorsed by science, the animal still manages to remind us of filth and infiltration, imagery that once made the rodent so perfect for anti-Jewish propaganda half a century before. With the War on Drugs in its second decade in the United States, we were told that the mind-altering effects of substances such as marijuana, cocaine, heroin and LSD were a threat to our safety, health and morality. The only defence we had against drugs was to say no in the first place. After all, it only took one taste and you were addicted. Rats proved it.

During the mid-twentieth century, animals pressing levers for rewards (or to avoid punishment) were often used to test how our own brains might respond to certain environmental or educational cues. Named after a renowned behavioural psychologist, so-called Skinner boxes seemed like the perfect way to study addiction. And as easy-to-raise mammals, rats seemed like the perfect subject.

Given the opportunity to self-administer happy-feeling chemicals, the rats in various drug studies went to town. This, ladies and gentlemen, is a rat's brain on drugs! And like the rat, when you start, you'll also have no choice but to succumb to your altered chemistry and use it ... and use it ... and use it ... until you're dead.

One man wasn't convinced. The Canadian psychologist Bruce Alexander suspected that even if drugs such as heroin and cocaine enticed our bodies to crave more, their influence on our compulsion was more a reflection of our general wellbeing than of the chemical's overpowering influence on our decision-making.

Drug addiction goes hand-in-hand with loss – not only of control, but of status, wealth, health and dignity. When most of us think of a stereotypical addict, successful surgeons don't come to mind. Nor do priests, heads of state or Supreme Court judges – we think of poor, ill people who have fallen from grace. Alexander wondered if we had it all wrong. Assuming the stereotype had any merit – which itself is contestable – what if those who were 'down and out' were instead more susceptible to embracing and maintaining habits that gave them pleasure?

The tide of combatants returning from Vietnam convinced Alexander he was onto something. A widely cited survey of soldiers revealed a relatively marked increase in their use of drugs, especially marijuana and opium.[1] While only 6.3 per cent of newly enlisted combatants confessed to being users of opiates prior to their tour, within months that number would spring up to 17.4 per cent. A later investigation published in *Archives of General Psychiatry* found that a fifth of troops who enlisted after 1970 were addicted to heroin at some time during their tour.[2]

Statistics like those mightn't be all that surprising. For one thing, in the hellish theatre of war, drugs provided a welcome escape, if not a superhuman boost when you needed to stay awake and aware. And the usual social taboos of 'normal' society relax when there are bigger things to worry about, such as surviving another day. Lastly, drugs such as opium and marijuana were freely available in the pre-industrial tropics of South-East Asia. They were said to be easier to get than alcohol.

But that wasn't the baffling part. Counter to expectations, 95 per cent of addicted veterans were found to have dropped the habit within ten months of returning to the United States. Completely cold turkey. Of those, only a handful touched heroin again within the next three years.

Even with the greater challenge of scoring illicit drugs in their local neighbourhoods, more than 5 per cent of returned veterans should still have been using regularly after they came home, right? After all, studies on addicts typically determine that around 80 per cent to 90 per cent pick up the habit again at some point, after even modern rehabilitation treatments.[3] Why were these soldiers so different? Alexander wondered how much of our addiction came down to the social environment, and how much was written in our biology and chemistry. He decided to put it to the test.

They referred to it as 'Rat Park', but in the numerous experiments Alexander conducted in the late 1970s at Simon Fraser University, in British Columbia, the large plywood box housing the test subjects was referred to as 'P'. It was a vast space, hundreds of times larger than your average laboratory rat cage, and held up to twenty rat 'citizens' at a time.[4] The happy little rats had it all: play equipment, food, space to socialise and mate in, and a choice of fresh water or water dosed with opium.

This contrasted with a set of steel cages, used as a control. The boring old boxes weren't necessarily bad for the rats – in fact, they were exactly the same ones most scientists used in any other study. Cramped and lacking stimulation, they also had a supply of drug-dosed water. These cages were Rodent Vietnam. Rat slums. Alexander coded them 'C'.

Rats coded with the letters CC would be weaned and raised in the steel cages for the whole experiment, which lasted eighty days. Likewise, those coded PP would live like 'aristoc-rats' in the park from the day they were born until the end of the experiment. Those coded with CP or PC were swapped from cage to park, or vice versa, at sixty-five days.

According to Alexander's results, rats raised in the steel cages took to the morphine like proverbial addicts. On the other hand, those raised in Rat Park showed no statistical significance in how much they drank from the opium-dosed water supply. Rats raised in the cages and brought into Rat Park tended to avoid the stronger opium doses, but did have a greater tendency to turn to the solutions when they were diluted and sweetened. It was almost as if they wanted the kick without the socially debilitating effects.

Over the next few years Alexander used Rat Park in other studies. In one he created super-addicted rats – who were given only opium-water for their first fifty-seven days – before letting them loose in rat heaven. He noted that they preferred the pure water over their old fix, although they suffered a small period of withdrawal.

Alexander felt vindicated by the results of his experiment. 'It soon became absolutely clear to us that the earlier Skinner box experiments did not prove that morphine was irresistible to rats,' he concluded.[5] 'Rather, most of the consumption of rats isolated in a Skinner box was likely to be a response to isolation itself.'

The results didn't make much of a splash at the time, and the experiments lost funding after several years. Yet in the decades since, Alexander's work has come to influence the debate over the nature of addiction more than nearly any other study; they're often lauded as solid proof that addiction is significantly sociocultural rather than solely biochemical. In 2007 Alexander received Simon Fraser University's Nora and Ted Sterling Prize in Support of Controversy. These days, if you read nearly any news story that challenges our perception of addiction, you can guarantee Alexander's rat nirvana will come up.

There is a caveat. As interesting as the Rat Park studies were, attempts to replicate their results have been somewhat mixed,[6] suggesting that the impact of environment is perhaps a little more nuanced than Alexander concluded.

Either way, the experiments now frame a heated debate over the nature of addiction and the question of whether it constitutes a disease. Today, psychologists hover between Alexander's belief that a happy rat is a sober rat, and the idea that 'one hit and you're hooked'. Addiction isn't guaranteed – or even highly likely – after exposure to common illicit substances. But there is some complex biochemistry going on inside our nervous systems that complicates any social theory.

Central to the question of disease are ideals of responsibility and normality. If drugs are considered to do bad things to your health, is it the fault of the user for succumbing to their temptation? How much freedom does a drug user have while under a drug's influence? Is it preordained in the chemistry of the compound, or due to circumstance? Are some people hardwired for compulsion, or does it all come down to the environment we're born into?

Since time immemorial, we've licked, drunk, injected and chewed things that make us feel weird and change how we behave. From fermented grains to funky fungi, be it for purposes of pleasure or altered thinking, for spiritual or human connection, it seems we humans have always taken time out to get off our faces. We've woven it into our culture in the form of ceremonies that bind and heal, as rituals that unite, and as tests of humanity, trust and joy. We've embraced a rich variety of substances that inebriate, stupefy, dull, enhance, gratify, dazzle or depress us, and for a wide range of reasons.

Yet over the past two centuries we've also outlawed the making and taking of most mind-altering substances. In spite of their history as pharmaceuticals, drugs such as cocaine, heroin and cannabis have become associated with immoral, altered behaviours – which, not coincidentally, are often viewed as a vice of cultural or ethnic minorities.

One of the first anti-opium laws was passed in California in 1875, fuelled in no small part by the proliferation of Chinese-owned opium dens and rumours of women being lured inside to their moral ruin. Nearly four decades later the United States passed its first law prohibiting the use of heroin. Cocaine had a similar story, being commonly used by African-Americans prior to (and following) their emancipation from slavery. Again, rumours of violence and rape precipitated the outlawing of cocaine in New Orleans early in the twentieth century. After the Depression,

Americans became suspicious and fearful of the growing tide of Mexican immigrants slipping across the border, bringing with them a drug few had heard of at the time – marijuana. The head of the Federal Bureau of Narcotics, Harry Anslinger, described cannabis as 'the most violence-causing drug in the history of mankind' – a claim that confuses those of us who have partaken of a joint or two in our time. So in 1937 the *Marijuana Tax Act* was passed, making it illegal to possess or transfer the drug for non-medical and non-industrial use.

Many addicts of narcotics or opioids fall into their habit as consumers of medically prescribed pain treatments. After all, regular use of a drug is less stigmatised when it's seen as medication, being authorised by a need to 'fix' something that is broken. In 2015, an estimated 2 million Americans were addicted to opioid painkillers,[7] with just under a third abusing their prescription.[8] Restricting access by cancelling or changing prescriptions can often lead to a search for an alternative supply. Without a socially acceptable need for the drug, a percentage will turn to an unregulated system to satisfy their cravings, with an all-too-often fatal outcome. Roughly one in twenty people who abuse their prescription go on to use heroin; four out of five heroin addicts in the United States claim to have started with prescriptions to medications such as oxycodone. Just over 33,000 people died in 2015 by overdose from prescription opioids, heroin and fentanyl.[9] That's 115 deaths in the United States every day – four times the number in 1999, and ten times the level in 1971, when the War on Drugs began.

What all of these statistics add up to is a problem in how we see addiction. Pharmaceutical companies can shoulder some of the responsibility, having reassured the medical community for decades that the risks of addiction were low. Our general impression of addiction, especially the way we question whether or not it is a disease, hasn't exactly helped matters either.

One of the first drugs to be criminalised is one we now take for granted as socially acceptable, in spite of its significant social toll. Anslinger could easily have been talking about alcohol as the most violence-causing drug in the history of humanity; that would have been more accurate.

The United States' experiment with alcohol prohibition was short-lived but infamous, inspiring a social backlash, as well as a proliferation of underground bars and smuggling networks. From 1920 to 1933, the distribution of most kinds of alcohol was illegal – with the exception of whisky, which could still be purchased with a medical prescription. Numerous other countries around the world have restricted or banned the sale of alcohol at various times and places, both widespread and small-scale. Today, leaving aside certain Islamic theocracies and selective bans in Indigenous communities where it's deemed a problem, alcohol remains freely available, the drug of choice for most of the world's adult citizens.

So it's impossible to talk about drugs and their addictive qualities without looking at how we judge the culture and behaviours of those who use them. The history of drug use in the twentieth century is complicated by cultural vilification, racial biases, criminalisation and spurious information. A person can feed their addiction to nicotine or caffeine during their morning break in ways a heroin or meth addict can't. Tallying up the personal and social harms of different drugs, alcohol comes out on top, followed – after a sizeable gap – by heroin and crack cocaine.[10] Yet it remains far more acceptable in many societies to say you were blitzed on Friday night and got into a fight than it is to say you shot up and chilled out.

Compounding this bias is the problem we face in identifying when an enjoyable habit becomes a troublesome compulsion that we struggle to control. It's not easy to objectively measure when a desire becomes a need, and when a need becomes a harmful addiction.

I grew up with an alcoholic father. All too often I had it explained to me that alcohol addiction was a disease, usually in justification of a relapse or a particularly severe bender. These days I understand he was asking for sympathy and understanding, but back then all I wanted was accountability. Somewhere in there was a man who I believed could choose to drink or not to drink.

To abuse or not to abuse. To control his rage, to show affection to his family. Yet to him, a force beyond his control was guiding his hand, unleashing a monster he could disown with the phrase 'it's a disease'.

My father thought he was the rat in the steel cage, desperately pressing that lever. We even moved to Rat Park when I started high school, a suburban oasis with a swimming pool and several acres of open space, in the vain hope it would make a difference. It didn't. I've been told he's sober these days, but the damage had been done.

I've long struggled with the question of whether his addiction was a disease, as he said it was. The question is one that has preoccupied medical and popular media circles for decades, and dates back a couple of centuries. In 1784 the American physician Benjamin Rush wrote, in his book *An Inquiry into the Effects of Ardent Spirits upon the Human Body and Mind*, 'Nearly all diseases have their predisposing causes. The same thing may be said of the intemperate use of distilled spirits.' Rush then listed various conditions that he believed made an individual vulnerable to addiction, including enduring long periods between meals, anxiety over health, professions that tax the mind (from which they seek relief through alcohol) and being under pressure or suffering disappointment.

In 1849 the Swedish physician Magnus Huss finally gave this alleged disease a name – *alcoholismus chronicus* – and provided the first instance of the term *alcoholism*. 'These symptoms are formed in such a particular way that they form a disease group in themselves and thus merit being designated and described as a definite disease,' he wrote, unequivocally agreeing with my father that addiction to alcohol is a flaw based in biology, not morality.

However Huss was obliged to point out that there is no observable 'break' in physiology: the symptoms weren't 'immediately connected with any of those [organic] modifications of the central peripheral portions of the nervous system which may be detected during life, or discovered after death by ocular inspection', he wrote.

The belief that alcoholism *was* somehow a physical 'modification' of the nervous system would come to form the foundation of that most famous of all self-help groups, Alcoholics Anonymous, which based its principles on the premise that 'alcoholism is a sickness, not a moral delinquency'.

Elvin Morton Jellinek, a New York–born author who in 1960 published a book titled *The Disease Concept of Alcoholism*, popularised the disease-based perspective under the funding and guidance of an early member of AA named Marty Mann. Nevertheless, he readily understood the challenge of marrying the compulsion to drink with sickness. 'Alcoholism has too many definitions,' Jellinek said, 'and disease has none.' To this unqualified outsider to the medical field, a disease was traditionally whatever the medical profession claimed one to be. And for many in the profession, given that *alcoholismus chronicus* lacked a clearly defined material basis, it just didn't qualify.

Jellinek attempted to overturn this view by surveying members of AA, categorising them along a continuum of dependency from alpha (an early-stage 'problem drinker') to epsilon (chronic dipsomania). Although his study was far from scientific, the structured classification helped give the rationale behind alcohol addiction an appearance of rigour. Four years before his book's publication, the American Medical Association had officially shifted its terminology – alcoholism was no longer an illness but a disease – though this was not without great controversy.

The fight to show that alcoholism – not to mention all other forms of substance addiction – is somehow predestined in biology draws on the belief that genes play a significant role in our physiological development. There's no shortage of studies over the decades, many including twins and adopted siblings, which indicate that predispositions to craving alcohol lie in our DNA. I for one grew up wondering about, and even fearing, my own predisposition to alcohol dependence, a worry that saw me go teetotal for some time.

Studies of neurological patterns have searched for structures and connections that hint at greater susceptibilities to the pleasure induced by various substances, or a reduced capacity to control our drive to seek ever-increasing degrees of satisfaction. Our hormones, genes, nerves, receptors and neurotransmitters have all been suspected of providing a material basis for a level of desire for a substance or habit that society says is morally unacceptable by changing our physiology in ways that increase pleasure, affect decision-making processes or intensify the risk of discomfort on withdrawal.

Colourful maps of the brain and genetic codes are today's germs. They are the physical flaw that identifies the undesired behaviour as a disease instead of a vice or sin. When addiction is a disease, there is medical intervention instead of judgement. Sympathy instead of blame. When it isn't … well, we're just weak and selfish.

In 2017 the famous Hollywood producer Harvey Weinstein faced accusations of subjecting dozens of women to unwanted sexual advances, harassment and degrees of sexual assault, including rape. He was subsequently reported to have checked himself into a clinic for sex addiction, feeding a bevy of opinion pieces and science columns to critically ask if Weinstein – or anybody, for that matter – can really claim that a sexually predatory nature is the result of an 'addiction'. At stake was his guilt: Weinstein was seeking to mitigate it, while the community at large wanted to maintain it. The discussion turned to hard biology in a search for answers, with experts comparing normal and abnormal levels of neurotransmitters such as dopamine, and drawing lines in the sand between average and pathological in order to weigh judgement against or in favour of 'diseased'.

Practically speaking, the question of whether addiction is a disease is an important one that goes beyond semantics. A neurologist and ex-addict named Marc Lewis, author of *Memoirs of an Addicted Brain* and *The Biology of Desire: Why Addiction is Not a Disease*, makes it clear which side of the fence he stands on. 'Will has an awful lot to do with it,' Lewis points out in an interview with *Vice* magazine.[11] 'A lot of addiction experts feel that self-empowerment, self-motivation, self-directed activities, self-designed goals for [addicts'] own progress are critical steps on the road of overcoming addiction. The medical model says you're a patient and you have to do what the doctor tells you.'

Lewis argues that defining addicts as patients who suffer a disease makes them passive; like a cancer patient lying back and letting the chemo do the work. By claiming that his addiction was a disease, my father was resigning himself to fate. There were ways to overcome his disposition, but those were in the hands of the medical authorities we visited at two in the morning when – sweating and shaking with detox tremors – he finally decided to seek help. To fix a disease, we are inclined to find and remove an organic cause, or dose it with chemicals that drag the patient back into an acceptable zone of normality.

Similar to Lewis, the American journalist (and also a self-proclaimed ex-addict) Maia Szalavitz prefers to describe addiction as a learning disorder rather than broadly label it a disease, even if she agrees that the distinction is somewhat academic. In her book *Unbroken Brain: A Revolutionary New Way of Understanding Addiction*, she compares addiction to developmental challenges. In an interview with Jennifer Ouellette at *Gizmodo* she explained: 'For me, very early in recovery, it was important to realize that I did not choose what happened. But I also understood that I did make choices over the course of my addiction.'[12]

This is the dichotomy of disease. We have limited control over whether our basic biology is broken or functional, but does that include things like willpower and our agency? Or is free will also subject to breaks in biology? Where in chemistry does the body begin and the 'I' end?

When it comes to the brain, there is a hazy disconnect between the functions of the mind and any processes external to our consciousness. If we're to agree that people like my father and Harvey Weinstein have a disease, we'd have to see their compulsion as solely the product of nerves and neurotransmitters behaving badly because they have the wrong kind of genes, or at least because they've had some mishap earlier in life. We'd need to accept that the choices they made were made harder by biology.

Yet what is agency, if not a product of the brain? It's hard to deal with blame and responsibility in an objective, material universe that has no place for miasmas and ghosts. Are we all rats in steel cages, or should we consider Rat Park open for business?

# BAD

There is a city inside each of our skulls. It consists of 100 billion or so tiny homes in numerous districts and suburbs, bifurcated by a chasm that divides it into a left and a right hemisphere. Deep in the middle, like the remnant of some medieval Old Town, are structures that govern our vital administrative processes. They're surrounded by zones of bureaucracy that interpret sensations, generate thought and emotion, and control movement: there are six lobes, each with their own responsibilities over our perception, reasoning and behaviour. Each is separated by fissures, distinct yet in constant contact via multiple neural autobahns.

A Swiss psychologist named Johann Gottlieb Burckhardt discovered it was possible to sever some of these highways by creating a physical 'ditch' between problem suburbs and the parts of our brain they were thought to affect, changing how an errant city functioned. In 1888 in the psychiatric clinic of Préfargier, in Marin, Switzerland, Burckhardt opened the heads of half a dozen patients in his care – lack of surgical experience be damned! – and removed entire neighbourhoods from the parietal, temporal and frontal lobes.

Most of these patients had been diagnosed with '*primäre Verrücktheï*', a condition that probably would be called schizophrenia today. They all displayed symptoms including delusion, aggression and auditory hallucinations. Burckhardt proudly claimed that half of his operations had succeeded, since three of the six patients showed a reduction in symptoms over the following months. Two of them technically no longer experienced hallucinations or agitation, but developed auditory verbal agnosia and so could no longer understand words. Two of the six developed epilepsy, which within a week contributed to the death of one. One patient remained abusive to the staff and became increasingly verbal. Another patient failed to improve at all, and developed speech difficulties, which led to a second operation.[13] Burckhardt ultimately counted that patient as one of his successes.

Burckhardt was philosophical about his work, claiming that he preferred 'it's better to do something than nothing'[Δ] to the traditional medical axiom of 'do no harm'. Within three years, and following much ridicule from his fellow physicians, the doctor followed a new axiom – quit while you're behind – and ended his research into what soon came to be called 'psychosurgery'.

The brain has been regarded as the seat of human behaviour as far back as the sixth century BCE. But it was really only in the seventeenth century that anatomists started to give this delicate lump of wobbly white meat a closer look, when a British physician named Thomas Willis sparked interest in the functions of the brain with his book *Cerebri anatome*. He also gave us the word *neurology*.

Despite all we have learned since, neurology remains a rather mysterious science. We know how muscles contract and relax to move our body, how the stomach squeezes and secretes acid to aid digestion, and how gases move across the lining of the lungs to give us oxygen. We have a basic grasp of how the pancreas releases insulin, which allows cells to absorb glucose, and how the kidneys filter salts to balance our electrolytes. But while we can map the connections and measure the chemical flashing of our neurons, on some level the precise workings of our brain still pose the same question scientists faced prior to the medical revolution of the nineteenth century – is the universe all physics, or is it possible that some functions of the human nervous system simply cannot be predictably modelled by numbers and classical laws?

---

Δ   The full quote is: 'Doctors are different by nature. One kind adheres to the old principle: first, do no harm (*primum non nocere*); the other one says: it is better to do something than do nothing (*melius anceps remedium quam nullum*). I certainly belong to the second category.' (See Manjila, S. et al. (2008) Modern Psychosurgery before Egas Moniz: A Tribute to Gottlieb Burckhardt. *Neurosurgery Focus*, vol. 25 (1).)

The early-seventeenth-century French philosopher René Descartes was one of the first to ponder the relationship between the intangible experience we call a 'mind' and the solid, meaty substances we can poke with a stick. On one hand, there's clearly a link between our brain and our ability to reason, imagine and feel. Damage to the physical tissue can change our memory and personality in significant ways. Yet based on Descartes' system of logic – which reduced everything down to the fundamentals of God and the fact that one's thoughts demanded one's own existence – the mind (or 'soul', since they basically amounted to the same thing) was philosophically a thing that God could make exist on its own, separate from the chemistry and physics making up our universe ... *if* He so chose.

The real question was how seemingly immaterial things, such as thoughts and free will, engaged with our bodies. Descartes figured that the dual realms of mind and meat somehow mixed within a nub of tissue called the pineal gland.

A century after Descartes, the German physician Franz Joseph Gall suggested that the mind controlled our body by engaging broad sections of the brain rather than by channelling actions through a tiny nub of tissue. Gall had little to say on the matter of the soul itself, choosing to leave the mind's immaterial side as an unknowable black box. Whatever it happened to be made of, the mind's influence literally shaped the physiology of our grey matter.

Described as the 'doctrine of the skull' (in German, *Schädellehre*), Gall divided the brain into twenty-seven organs, which were demarcated by measured deviations in the smoothness of the cranial bones. Each of these organs was responsible for a faculty of the soul, such as facial recognition, motor skills or colour sense. There were also specific faculties for poetic talent and the desire to commit murder.

In his book *The Anatomy and Physiology of the Nervous System in General*,[Δ] Gall outlined the reasoning behind his theory. For the

---

[Δ] Or, to give its full title, *The Anatomy and Physiology of the Nervous System in General, and of the Brain in Particular, with Observations upon the possibility of ascertaining the several Intellectual and Moral Dispositions of Man and Animal, by the configuration of their Heads.*

first time in history, an attempt was made to detail links between the brain's physiologies and precise functions. Measuring the bumps on a person's skull according to the applied science of cranioscopy would allow a physician to diagnose personality traits.

If you think this sounds more familiar as the pseudoscience of phrenology, you can thank the physician Johann Spurzheim for jumping on board and selling the idea to the Brits in 1813 under the new name. While the pair collaborated at first, Gall was none too happy with the partnership as time wore on, accusing Spurzheim of stealing his work and making a complete mess of it.

Regardless, Gall's idea took off, in particular appealing to the industrial middle and upper classes in an early form of pop psychology. Interest in science was a sign of class and worldliness, and people were increasingly keen to understand themselves in an academic light. It was one thing to suspect a neighbour of deceit or to declare one's own talent for the arts; bumpy noggins could now prove the details of human nature at a touch. People consulted phrenologists on everything from marriage compatibility to tailoring education for children to weighing the guilt of a defendant.

Cranioscopy and phrenology were variations of physiognomy – the art of using anatomical traits to interpret a person's character. We've attempted to connect facial features with personalities since before the days of Aristotle, who himself claimed that having a broad face displayed lower intelligence, while a small face showed one to be resolute. Even today we still use the face as a symbol for personality, giving overbites and weak jaws to comical fools or broad chins to heroes, and use terms like *lowbrow* and *beady-eyed* to imply class or trust. The Swiss philosopher Johann Kaspar Lavater reintroduced the practice to the public in the late eighteenth century in a book beautifully illustrated with woodcuts of caricature profiles, paving the way for over a century of scientifically sanctioned judgement, while also giving us terms like *stuck-up* for conceit, and *profile* for an assessment of character. Physiognomy lent scientific credibility to the idea that you could judge a book by its cover.

By 1840, phrenology was all but discredited as advocates struggled to agree on the exact numbers and locations of specific brainy bits. Damningly, to the conservative religious members of the community, phrenology left too little room for a soul, and so became associated with atheism, and therefore with immorality.

So that was that. Mostly. There were occasional revivals of the system through the twentieth century. The so-called father of criminology, Cesare Lombroso, claimed in 1903 that the shape of a man's skull was evidence of his tendency to break the law. On autopsying the brain of a convicted murderer in 1871, he wrote:

> I seemed to see all at once, standing out clearly illuminated as in a vast plain under a flaming sky, the problem of the nature of the criminal, who reproduces in civilised times characteristics, not only of primitive savages, but of still lower types as far back as the carnivora.[14]

Lombroso didn't stop at skulls, either. 'What is sure is that criminals are more often left-handed than honest men,' he claimed, 'and lunatics are more sensitively left-sided than either of the other two.'[15] As late as the 1930s, Belgian colonists in Rwanda were implementing national phrenology-based programs to justify their discrimination between the ethnic identities of the Hutu and Tutsi people.

Phrenology coincided with dramatic advances in our understanding of neurology. In the late eighteenth century, the Italian physician Luigi Galvani famously skinned a frog and brushed its sciatic nerve with a scalpel, making its leg twitch and sparking a line of inquiry that would unite electrochemistry with nervous action. In the middle of the nineteenth century, physicians such as the French doctor Paul Broca linked physical damage to localities in the brain with changes in behaviour and abnormalities, building cerebral maps in ways that both reflected and contrasted with Gall's doctrines.

By the time Burckhardt began to chisel windows into his patients' skulls, the medical establishment was reconciling evidence of the brain's localisation of certain behaviours, such as speech, with a belief that a person's psyche was hazily diffused throughout the whole organ. Burckhardt was a proponent of the brain as a modular system – and as such, he believed it was possible to manipulate parts of it to 'fix' certain problems.

His legacy was mud across the medical community, but it didn't take long before another physician was tempted to apply the scalpel to brain matter in order to repair the mind. The Portuguese neurologist António Egas Moniz was reportedly encouraged by a 1935 conference presentation delivered by American neurologist John Fulton on his chimpanzee vivisection research. Fulton claimed that severing key nerves linking the frontal lobe with the rest of the brain had pacified his two chimpanzee subjects, to the degree that it was as if they'd joined a 'happiness cult'. By one account, Moniz asked during the Q&A if the practice might be applied to humans.[16]

For his part, Moniz claimed he came up with the idea long before the conference. The preceding decades had in fact unveiled a great deal about the functions of the frontal lobe as a 'seat of reasoning' – the lobe responsible for much of what makes us human. World War I had provided researchers with no end of brain-damaged subjects to study, after all.

Regardless of where his inspiration came from, in 1935, in Lisbon's Hospital Santa Marta, Moniz directed a series of operations on mental patients intended to destroy the fibrous highways between the frontal lobe and the brain by injecting alcohol through a small window cut into the skull. By February the following year, Moniz and his team of surgeons had operated on twenty patients with diagnoses including schizophrenia, depression, mania and panic disorders. According to his own notes on his patients, complications were transitory and included 'increased temperature, vomiting, bladder and bowel incontinence, diarrhoea, and ocular affections such as ptosis and nystagmus, as well as psychological effects such as apathy, akinesia, lethargy, timing and local disorientation, kleptomania, and abnormal sensations of hunger'. He also claimed that two-thirds of his procedures were successful (one-third significantly so), with the remaining third unchanged. It should also be noted that none of the patients died, giving Moniz something else to celebrate – along with his 1949 Nobel Prize in Medicine.

Not everybody was so overjoyed with Moniz's accolades. As with Burckhardt, the medical community was more outraged than impressed. A physician who had supplied Moniz with a number of patients believed the reported changes mostly came down to general brain trauma and shock, rather than to any precise effect of the leucotomy itself. He called Moniz's theory 'cerebral mythology'.

Not that it mattered – whether down to Moniz's self-promotion or to optimism regarding this brave new field of medicine, psychosurgery was soon being performed by a number of surgeons across the world. Neurologists from Romania to Brazil experimented with new ways to reach deep into the brain and transform the frontal lobe's real estate. An Italian psychiatrist by the name of Amarro Fiamberti found that a more efficient pathway to the connecting fibres was through the eye socket, eliminating the need to cut large holes into the skull.

One physician on whom Moniz made a particular impression was an American neuropsychiatrist by the name of Walter Freeman. After a chance meeting in London at a neurology conference, the pair began a correspondence, which ultimately gave Freeman the courage to make the jump into psychosurgery.

Freeman and his neurosurgeon colleague James Watts developed a technique called the 'precision method', in which a blunt spatula was used to sever the fibres through an opening on the side of the skull. Yet Freeman felt the process was still too restrictive, requiring an operating theatre and the sterility that went with it. Fiamberti's transorbital method was just what he needed for a less invasive, more mobile procedure, one he could take on the road, and so not be anchored to hospitals. His reasoning was more or less altruistic, understanding that those who stood to benefit most from the procedure weren't wealthy enough to afford the costs of surgery, and often couldn't travel far.

Freeman performed the first of these 'frontal lobotomies' in 1946. The act more resembled medieval torture than a medical procedure, and was done relatively blind. You might want to skip the next paragraph if you're squeamish.

An ice-pick-like implement called an orbitoclast was slid under the eyelid and over the eyeball, then lightly hammered through the thin bone at the rear of the socket. Having aligned the tool parallel with the bridge of the nose and at a slight upward tilt, Freeman would slide it several centimetres into the frontal lobe, before angling it towards the nose and back again. It was then inserted a little deeper before again being swished back and forth to damage the necessary nerves. Don't try this at home, kids.

Moniz was none too happy with this new kitchen table method of leucotomy. Neither was Watts, who in 1947 dissolved his partnership with Freeman and went his own way. In any case, the age of the frontal lobotomy had arrived. In 1949 alone, more than 5000 leucotomy procedures were performed in the United States, and there would be some 40,000 similar operations over the next several decades. Thousands more were performed across Europe.

Critics saw the procedure as barbaric and brutal, far too imprecise to be useful for delicate adjustments to personalities, and believed that it risked side-effects that were often worse than the original condition. Many were ideologically offended, arguing that it reduced the mind to simple modules which could be switched on and off, in contrast to theories that saw personality as a diffuse system which emerged from the brain as a highly interconnected entity.

On the other hand, those in favour of leucotomies cited testimonies from the satisfied families of patients, and research that emphasised a significant success rate. A study conducted in 1947 called the Columbia-Greystone Project[17] 'volunteered' forty-eight subjects who had been institutionalised for at least two years and who had demonstrated no recognisable signs of improvement. Half received a topectomy – a procedure not unlike a leucotomy, in which tissue was removed from the prefrontal cortex – while the remaining half were left untouched. Psychiatrists evaluated the forty-eight patients four months later, not knowing who had received the surgery. The study concluded that the topectomy didn't affect patients' learning or general neurological behaviour, and even reduced their anxiety and improved their 'mental attitudes'. Nearly half the subjects were paroled shortly after; most came from the group that had surgical treatment.

So who was right? With the benefit of hindsight, both were. Psychosurgery typically altered each patient's personality in ways that reduced or eliminated outward signs of heightened emotions, whether they were morose, violent or manic. Unfortunately, such changes were rarely limited to the undesirable traits, and so patients often exchanged one form of suffering for another.

If there's a textbook case study that perfectly frames the tragedy of the lobotomy, it would be the story of Rosemary Kennedy.

The Kennedys just might have been twentieth-century America's royal family. Father Joseph P. Kennedy Snr was a well-known businessman and ambassador to the United Kingdom's Court of Saint James. Rosemary's older brother, John, was the thirty-fifth president of the United States. Two of her younger siblings, Bobby and Ted, would grow up to become senators. Rosemary was raised in a world of power, politics and class, where poise and intelligence were more than virtues – they were integral to the family's identity and reputation.

Rosemary's complicated birth, rumoured to have been caused by her mother's concentrated efforts to abstain from delivery until the doctor arrived, contributed to her developing intellectual delays as a child. In spite of extra tuition and private schooling, the young child failed to progress past fourth grade, and was evaluated to have an IQ of roughly sixty to seventy. Her family went to great pains to hide her intellectual challenges from the world, refraining from discussing her limited ability to read, write and count with her teachers and tutors. Mental disabilities of any kind were viewed with general discomfort and shame in most areas of early-twentieth-century society, yet for gentry like the Kennedys such intellectual shortcomings were an outright embarrassment.

Rosemary was painfully aware of how her academic skills compared with those of her eight brothers and sisters, although by her own account her childhood in the quiet English countryside was still largely a happy one.

An easy-going adolescent, by her early twenties Rosemary began exhibiting violent outbursts, raging against the restrictions her siblings had escaped, and by some reports experiencing fits that resembled those caused by epilepsy. In 1940 Rosemary's father resigned as ambassador and the family returned to the United States. Her wild mood swings only worsened, and her family grew desperate for a solution.

That fix came in the form of Freeman and his assistant, Dr Watts. In 1941, five years before Freeman took his orbitoclast to the streets, Rosemary Kennedy had her head shaved and her frontal lobe disconnected. The results were a complete success ... and an abysmal failure. The young woman was reportedly calmed of her 'rages'. But she lost her speech, her ability to walk and her bladder control for several months following the procedure. Even then, Rosemary forever walked with a limp and had a speech impediment. Her mental abilities went from those of a young adolescent to those of a small child.

Rosemary was sent to a private psychiatric hospital in New York, then later was relocated to another facility in Wisconsin, where she spent the remainder of her life, completely removed from her family. Her own mother did not visit her for twenty years. Joseph never visited her at all.

If any good can be said to have come from the tragic consequences of Rosemary's treatment, it would be her powerful family's eventual rallying to the rights of the mentally disabled. By the 1960s her mother and her sister Eunice spoke openly about mental health, while in his presidency John advanced legislation supporting the rights of those with intellectual challenges. In language considered progressive at the time, he stated: 'Mental retardation ranks with mental health as a major health, social, and economic problem in this country. It strikes our most precious asset, our children.'

Rosemary's family sought to cure what they saw as a broken young woman who didn't meet their expectations of intellect, discretion and cooperation. Shame, sympathy, frustration, isolation from help ... many things no doubt drove their desperation for a fix, one promised by surgeons who, like Burckhardt, figured it was 'better to do something than nothing'.

There's no doubt Rose and Joseph loved their daughter dearly; by all accounts, both were wracked with guilt over what they'd done to their child. Even Rosemary herself would probably have judged her body's functions – her difficulty in learning as quickly as her siblings, her drive to lash out, and perhaps her epileptic fits – as abnormal and undesired by comparison with those around her.

The real tragedy was not that surgeons operated on Rosemary's brain. Neurosurgeons routinely remove lesions and tumours with little judgement. Even cutting the connections between lobes and hemispheres is rarely seen as an offensive act when it saves the life of somebody with a severe form of epilepsy. The lobotomy's curse was one of self-limiting options. Rosemary's brain was a wild garden that her family believed needed to be pruned into submission, rather than accepted for its own beauty. Culture had framed low intellect and heightened emotion as a disease that Freeman should cut out. For all their privilege and wealth, Rose and Joseph found few opportunities in their community to help them adapt to their daughter's variations in neurobiology. They couldn't beg her to change, bribe her to improve or punish her into compliance. It wasn't her fault. The family looked to change her physiology instead.

Just as the competing theories of contagions and miasmas were contrasts between the tangible and the nebulous, the dualism of the mind and neurology was effectively a dichotomy of an opaque system of free choice and the cold, deterministic mechanics we call physics. From Gall to Burckhardt, Descartes to Freeman, physicians and philosophers have long been desperate to reconcile the complex nature of offensive thoughts and actions with deviations in tissue and organ, a nature we can be tempted to fix with a scalpel or a chemical.

When a sin is the product of illness, we have more to offer than pity: there is permission to physically interfere with the city inside a person's skull, righting broken rules in order to return them to an acceptable level of virtue. On the other side, when bad behaviours are judged to be the product of a free mind, there is more than shame: there is permission to exact justice, compensation and retribution.

# PSYCHO

In one of my favourite historical artworks, the eighteenth-century English artist William Hogarth shows the rise and fall of a fictitious wealthy merchant's heir. Tom Rakewell may be one of the most pitiful wretches never to have existed. The first seven frames show him spending his inheritance on various vices, descending step by step from grace. In the eighth and final scene, a fallen, half-naked Tom is ogled by wealthy socialites touring the notorious Bethlehem Hospital in London. *A Rake's Progress* is a warning against deviance, framed by a 'rakehell's' imprisonment for wanton immorality, swiftly followed by institutionalisation in a mental asylum.

It's a common theme in art and literature – the dark twins of evil and madness, each commonly taken for the other, both treated with exorcism, ostracism and exile. Mental illness and criminality are separated by the thinnest of borders, one that threatens to vanish if examined too closely. From haunted houses to horror movies, Hogarth paintings to Batman comics, the trope of the criminally insane is often used to excite fear arising from an absence of reason and predictability.

This distinction is more than just a device for fiction. Framing the actions of a criminal in terms of hardwired neurology can change how a jury judges their offence, for example. Evidence used to defend a mentally ill offender at a trial is more likely to be successful if it is described in terms of biology than psychology.[18] To be a psychopath is to sit on the edge of madness and badness; it's a medical label not just for a physical truth but also to serve as a warning that the afflicted lacks the all-important sense of empathy we rely on to form trusting relationships. When the prosecutor, rather than trying to disprove mental illness, describes the offender as a psychopath, juries are slightly more likely to be persuaded by their argument.

Our desperate need to reconcile crime with mental health stretches the definition of disease beyond breaking point. What was a convenient, if flawed, model for treating suffering becomes a lodestone around our necks as we do our best to weigh mutually exclusive concepts of pity against blame, treatment against retribution, and public safety against body autonomy.

I remember watching news footage of the carnage left behind by the far-right Norwegian terrorist Anders Behring Breivik. It's hard to know what to focus on when you hear that a single man has taken the lives of seventy-seven people – was he inspired by extreme political ideology, extreme evil or extreme neurology? Nothing seems satisfactory in the face of such grief and fear; each explanation means we must give up hope, understanding or justice.

Breivik is a sane man in the eyes of the law. His decision to take the lives of as many people as possible on 22 July 2011 in Oslo, Norway, was concluded by a court to be the product of a free, if starkly evil, mind. Breivik was found morally responsible for those deaths, and by order of the court won't walk the streets again until at least 2033.[19] At the time of writing, he remains in isolation in a Norwegian high-security prison. But it could have been a rather different story.

Prior to Breivik's trial, a psychiatric assessment conducted by court-appointed psychiatrists concluded that the defendant exhibited symptoms of paranoid schizophrenia. They said that Breivik lived in his 'own delusional universe where all his thoughts and acts are guided by his delusions'.[20] A second, independent assessment found no sign of such a condition, concluding that, at worst, Breivik was dissocial and had a narcissistic personality disorder. In other words, even if his psychology was a little flawed, it presented no reason to believe he was not morally responsible for his actions. A panel of five judges unanimously accepted the alternative conclusion, finding Breivik was of sound enough mind to be responsible for his actions.

Their verdict of guilty resulted in the mass murderer receiving the most severe sentence in Norway: 'preventive detention', or the containment of an individual without hope for their rehabilitation. In Breivik's case, that means a twenty-one-year prison sentence that can be extended indefinitely.

Beyond the world of wigs and suits and Latin phrases, and lacking the expert evaluations, most people were convinced beyond any doubt of his sanity. He had to be sane. A poll conducted by Norwegian public broadcaster NRK found that three out of every four citizens believed Breivik was mentally competent when he pulled the trigger, and should therefore go to prison if found guilty.[21] While psychiatric care would also have seen the murderer separated from the community, mental health facilities aren't the same as prisons. Prisons are where bad people go, and this, indubitably, was a bad man.

Emphasising the divide between mad and bad, Breivik himself considered the label of insanity to be worse than death. 'To send a political activist to a mental hospital is more sadistic and evil than to kill him!' he claimed.[22] His actions were a reflection of his ideology, and he desperately needed this ideology to be seen as the product of a sane, free, clear-thinking, functional and intentional mind. Being diagnosed with a mental disease would imply Breivik lacked a degree of control over his actions due to a flaw in his mental physiology, and therefore wasn't ultimately responsible.

And so both Breivik and the public wanted his atrocities to be accepted as an act of liberty. For Breivik, liberty lent credibility to his message as reasonable and righteous. For the public, liberty meant culpability, and that permitted retribution.

In justice, the tension between responsibility and mental illness dates back to Roman times, but the modern-day 'insanity defence' was generated in the wake of a famous trial, which followed an attempted assassination of the British prime minister (and the murder of a civil servant) in 1843. The House of Lords grilled a panel of judges over the case, which eventually led to the passing of a common law demanding that trials take into account the accused's mental state. Interestingly, the accused murderer and would-be assassin – one Daniel M'Naghten – wouldn't have passed the test that was subsequently established.

The original insanity defence states:

> [I]t must be clearly proved that, at the time of the committing of the act, the party accused was labouring under such a defect of reason, from disease of the mind, as not to know the nature and quality of the act he was doing; or, if he did know it, that he did not know that what he was doing was wrong.[23]

'Defects of reason' and 'disease of the mind' are fairly fuzzy concepts. The Commonwealth criminal code further qualifies its version of the defence by stating that mental impairment must be factually established by way of expert opinion, adding: '[M]ental illness is a reference to an underlying pathological infirmity of the mind, whether of long or short duration and whether permanent or temporary, but does not include a condition that results from the reaction of a healthy mind to extraordinary external stimuli.'[24]

By these rules, some kind of pre-existing brain trauma resulting in severe intellectual impairment might be considered reasonable grounds for such a plea. Identical behaviour resulting from a three-day binge on drugs and alcohol might be evidence of diminished responsibility, but it's unlikely your defence team would entertain it as an argument for insanity.

Keep in mind, I'm as much a lawyer as I am a doctor – which is to say, I'm not. So this isn't legal advice, just a demonstration of how thin the line is between disease and damnation. I promise you, nothing in this book will help you avoid time in the clink.

Not that a plea to take mental health into consideration is a common tactic. Not by a long shot. Mental illness might seem like a good way to escape a prison sentence, but in fact it's cited as a defence in barely 1 per cent of cases, and just a third to a quarter of those pleas of 'not guilty by reason of insanity' succeed.[25] Admittedly, some crimes, such as homicide, attract more insanity pleas than others, but overall it isn't the trump card we might take it for.[26] Most defendants want to avoid the stigmatisation of being seen as insane, but they probably also fear that a sentence in a mental institution might even be worse than one in a prison.

Such a fear isn't entirely unjustified. A report into mental health conducted in Australia by human rights commissioner Brian Burdekin in 1993 expressed concern that medicalised detention 'can be a particularly severe punishment because it is not subject to the normal legal protections which apply to those convicted of crimes'.[27] A mental health ward is not, strictly speaking, a penal institution, so patients aren't necessarily covered by the rules and regulations that, in a prison, might help ensure they receive fair representation or consensual treatment. Nor are the forensic hospitals themselves necessarily constructed with the same protections for patients and staff. Even with the state and national reforms that followed Burdekin's report, there remain fundamental differences in how we as a society see the autonomy, guilt and rights of criminals and patients.

Patients aren't prisoners. But our ability to distinguish between who is sick and who is sinful is at best flimsy, and at worst based more on our social biases than on anything medical. There is no pigment, no chemical, no gene, no physical kink or fracture that can delineate the insane from the free-thinking. Nothing on a brain scan will conveniently show us who is capable of knowing right from wrong, or of making their own decisions with an unfettered will.

Not that we haven't looked for a scientific solution to picking the disturbed from the diseased ...

# GUILTY

Ideally, the goal of a justice system is to determine not just a course of events that led to a harmful act, but the motivations of those who were behind that act. But hard evidence can only provide so much detail. Beyond a certain point, the desires and intentions of a defendant are forever out of reach, locked up in the complex weave of grey matter. Yet some neuroscientists think it's possible to unlock that box and look inside.

In 2016 a team of researchers scanned the brains of forty people as they took part in a role-playing exercise designed to play with their minds.[28] The volunteers could choose a suitcase at random to take through a virtual checkpoint. Easy enough. One of those bags, though, carried items that would earn them a reward – if they weren't caught by the guard.

At the start of each trial, half the participants were informed of the risk of getting caught with the goods at the checkpoint. They were then shown a number of bags and asked if they would like to choose one. Sometimes there was just one bag to choose; sometimes they could select from up to five. It stood to reason that the fewer bags they had in front of them, the more they knew they risked committing an act that would result in a loss.

Meanwhile, the other set of volunteers were shown the number of bags first, asked if they were willing to select one, and only then told the risk of getting caught. This made it more a game of Russian roulette than one of deviant plotting.

By running the resulting brain scans through a program capable of detecting subtle patterns, the researchers found they could predict fairly accurately who among the subjects was sweating it – thinking they had picked up an illicit bag – and who was fairly sure theirs was safe. The patterns were most obvious among those who intentionally picked from fewer bags while knowing there was a good chance of getting caught.

Putting it simply, the scans showed who knowingly committed a crime and who believed they were unintentionally breaking the law.

Studies like these aren't without their problems, and can't be taken as more than a nod at something that needs closer investigation. Still, it's potential evidence of a physical link between guilt and the brain, showing that our intentions do briefly leave fingerprints in the physical realm.

What the study doesn't do, however, is show how the criminal act itself might be predestined by biology. Are some people more inclined to become criminals because of their neurological wiring? A study conducted in 2013 attempted to answer this question by looking at how variations in brain structure might favour impulsive behaviours, which make breaking laws more likely.[29]

Ninety-six male inmates from two New Mexico prisons took part in a simple task: they watched a screen and pressed a button every time they saw the letter X. The letter would flash up on the screen 84 per cent of the time; the remaining 16 per cent of the time the letter K would flash up instead.

The experiment tested the premise that more impulsive inmates would find it harder to avoid pressing the button when they shouldn't. Brain scans of the offenders taken during the experiment showed decreased activation in an area of the brain called the anterior cingulate cortex, a chunk of grey matter that curves beneath the frontal lobe and is thought to help us detect mistakes. This correlation also matched the time it took for the offender to rescind once they were released from prison at a later date: about half of them were arrested again for a crime during the next three years.

Wonky anterior cingulate cortices are just one way the dice could be loaded for some unfortunate souls. On 1 August 1966 a man named Charles Whitman killed his mother and wife in a premeditated act of murder, then climbed a tower on the University of Texas campus and took the lives of fourteen random passers-by with a rifle. Whitman was shot dead, but a suicide note he left behind stated: 'I do not really understand myself these days.' He also requested that an autopsy be conducted on his body, to determine what was causing his odd behaviour and the headaches that had been plaguing him. Whitman was in fact found to have had a tumour on his brain.

The case has little scientific value on its own, but stands as a shocking example of the possibility that brain trauma could be behind some criminal actions. Studies on lesions and criminal activity have identified possible networks that might raise the risks of acting against our better moral inclinations, should they become impaired.[30]

If there is merit to such research, just how much should we view crimes as symptoms of disease? Brain damage is one thing, but what if we could show a 'criminal' neurological structure encoded in an offender's genes?

In 2014, 900 offenders in Finland had their genetic make-up analysed. The investigation revealed a pair of genes that could – according to the researchers – be a factor in 5 to 10 per cent of all violent crime in the country.[31] Just having those genes made someone thirteen times more likely to have a history of aggressive behaviour. One of the genes, a variant of cadherin 13, shows up frequently in those with ADHD and substance addictions.

The second, called MAO-A, sits on the X chromosome and encodes an enzyme called monoamine oxidase A. It has a mutant variant which can result in a condition called Brunner syndrome, characterised by impulsive behaviours and learning difficulties. Having a low-active form of the gene, dubbed MAOA-L, also correlated with aggression, earning it the nickname 'warrior gene' in the popular media.

The discovery of the MAO gene's relationship with anger and violence has invited speculation over whether such genes could have been shaped by evolution. MAOA-L is twice as likely to appear in New Zealand's Māori population as in its European population, for example, a fact that has fuelled stigma among the country's indigenous communities. One Māori scholar, Gary Hook, has suggested that 'because Māori evolved in a high-risk environment, survival favoured those mutations that contributed to his survival and hence the frequency of the "warrior" gene in the Māori population became enhanced over those found in other races'.[32]

What this means is that a gene that might impede an ability to meet social expectations in one culture – what we might consider a disease – might produce socially prized behaviours in another, leading to its proliferation.

Asks Hook: 'Could those Māori who express the "warrior gene" be diagnosed as having a medical condition similar to those with diseases such as Brunner syndrome or Norrie disease, two diseases involved in the expression of abnormal MAO genes?' To extend the question, could we consider all violent offenders who have abnormal MAO and/or cadherin 13 genes to have diseases?

Once again, we come back to the qualifier of blame, and the hazy junction between the material body and the black box of our mind. Considering those with problematic biology as having a disease means agreeing that they are less culpable, after all, prompting questions about our mind's relationship with our genes, our chemistry and the physical form of our nervous system.

If explaining how brains make decisions is hard, tracing the process all the way down to a point we can define as truly *free* is virtually impossible. To understand why it's a problem, we can consider the thought experiment of the 'philosophical zombie' – a robot that looks and behaves like a human in every way imaginable, but that is fully programmed according to a set of rules.

Instead of human brains, these fleshy robots have supercomplex computers, though of course we can't see them. Think of the perfect Siri, if you like, responding to your questions in such a way that you'd swear she was real. Artificial intelligence technology might eventually improve Siri so she'd fool more and more people, but would that mean Siri will one day become aware of her own thoughts? More importantly, at what point – if ever – will her decisions occur independently of her programming?

This question is referred to in philosophy as the hard problem of consciousness. It's virtually impossible to solve, since it involves coming up with a reasonable explanation for how a physical thing like a brain can reflect and think in a way we might agree isn't predetermined by genes, nervous connections or the chemistry of neurotransmitters.

Part of the problem of free will is that our understanding of everything in the universe falls into just two categories: things that are firmly set by a law, and things that can't ever be predicted. As any good snooker player knows, if you hit a ball in the same place with the same force, it will always go in the same direction. Every single time. That's determinism. It gives us rules of chemistry and engineering, and so helps us build skyscrapers, send robots to Mars, treat diabetes and bake biscuits that taste delicious every single time.

In the impossibly small landscape of quantum mechanics, physicists know that some things won't have a preceding cause that results in a specific outcome. In technical jargon, we say there are no local variables accounting for some events. The precise moment a radioactive atom spits out a particle has no cause, no local laws – it is *random* in the truest sense of the word – even if there is a general rule that tells us roughly how often we can expect it. The universe is truly random on a tiny, tiny scale.

So here's the catch. Guilt requires a person to have a mind that is free to choose between right and wrong. Since it's not anchored to one outcome, it's not predetermined by physical laws. After all, if a person couldn't make any other choice, could we blame them for being wrong?

So if something isn't determined by hardwired laws, it must be random. Frustratingly, that doesn't feel right either. It seems weird to think of our choices between right and wrong as just a flip of a perfectly balanced cosmic coin. A decision is ultimately caused by

'you' – that mysterious element beyond brain cells and DNA we think of as a person. It's not arbitrary.

Maybe free will is a mixture of quantum randomness and determinism. Perhaps, but even mixing rules and coin flips falls short of describing our experience of making a truly free choice between eating a hot fudge sundae or a blackforest cake, or between swiping a credit card and handing back a dropped wallet.

This isn't to say free will doesn't or can't exist (although some philosophers do come down on that side of the fence). It isn't to say we won't one day solve this hard problem (though some argue we won't). There are solutions, but none of them are scientifically solid.

What all this means is that we can't actually see or even agree upon the line we say distinguishes between the predetermined consequences of a disease and the freedom of criminality; what's more, the line itself has no universal definition. Regardless of where we stand individually on the question of free will and its implications for willpower, morality and responsibility, we're ill-equipped to lay the ground rules that will determine where physiology ends and unfettered choice begins.

Philosophers will be arguing about how to box our free will for a long time to come. Not that we could act any differently even if we did come up with a solid answer. For now, and perhaps forever, we will be left with the frustrating challenge of finding a way to distinguish between the functions of our brain we can claim responsibility over and those we can't – finding the border between insanity and evil, between diseased and culpable, between shame and pity.

# PERVERTED

I'm going to bet you've never come across Gregory George Glebow. I had to look his name up myself, to be honest, and his actions had a profound effect on my life nearly twenty years ago.

Just before starting work in a hospital pathology lab, I had a position in the organisation's central distributions centre. Specimens would be delivered into what was effectively a large sorting room, where a ragtag bunch of us grunts boxed and filed the blood, piss and cancer into trays for carrying to the various laboratory stations.

That's where I met Michael Greer. I can't say I knew him all that well, or for very long. But the lanky, scruffy, happy-go-lucky guy was so affable, so contagiously friendly, that I don't think anybody could meet him and not immediately consider him a mate. Since he lived a few doors down from the labs I occasionally hung out at his place, and we gravitated naturally towards each other during coffee breaks and at the sorting bench, talking music or gaming or general gossip.

Nine months later I'd moved on to greener pastures, and over the next few years lost touch with Michael. I don't remember exactly how I later heard he had died, whether through a mutual friend or a chance reading of a newspaper. But the event itself was a punch to the gut I'll never forget.

Walking home one night in March 2000, his arm draped around a friend's shoulders in platonic love or sympathy, Michael was assaulted by Gregory George Glebow. The attack was so ferocious that it left him in a coma, and he died months later of his extensive injuries.

What could have inspired such a heinous act? Glebow assumed his victim was gay.

Not that it's relevant, but I can't tell you much about Michael's sexual interests. He was a father to a child from an earlier relationship, and when I knew him I had one hell of a crush on his girlfriend, for what it's worth. But I'm certain Glebow didn't know much about Michael's sexuality either. Really, Michael's life ended because another man judged him to be too caring, too affectionate and perhaps too effeminate. That, in Glebow's mind, deserved punishment.

We can take some small comfort from the fact that Glebow is serving a life sentence for his unfathomable act of violence. Even while incarcerated, the man has continued to kill, taking another inmate's life in 2012. But we shouldn't take his incarceration for granted. Only recently, in 2017, did Queensland's parliament finally scrap a law that allowed a perceived homosexual advance to be grounds for assault, also known as the 'homosexual advance defence strategy'.[33] Shockingly, this is still a viable legal defence in South Australia. Justice in Michael's case was fortunate, even if it wasn't inevitable.

Societies around the world have been slow to repeal archaic laws that criminalise consensual sexual acts between men. Across different cultures and in many nations, including Libya, Nigeria, Sudan, Mauritania, Qatar, Iran and Saudi Arabia, male homosexuality remains an illegal act punishable by lengthy prison terms, corporal punishment or, for repeat offences, death. Fear and discrimination persists even in cultures that have otherwise repealed such laws.

The legality of homosexuality among women has a similarly chequered past. In many nations where sexual activity between men can earn anything from prison to whippings, such as Ghana, Jamaica, Turkmenistan, Kuwait and Malaysia, women face no such threat. Yet throughout history, similar edicts have explicitly warned women against engaging in sexual conduct with one another. A thirteenth-century French law, spelled out in the code *Li livres de justice et de plet*, states:

> He who has been proved to be a sodomite must lose his testicles. And if he does it a second time, he must lose his member. And if he does it a third time, he must be burned. A woman who does this shall lose her member each time, and on the third must be burned.

While overall our attitudes towards homosexuality are gradually becoming more progressive, to a number of communities around the world being gay is a condition that can – and should – be either cured, punished or suppressed. Antidiscrimination laws and marriage equality aside, people continue to be discriminated against for their sexual orientation and practices across the social spectrum.

The fact that attitudes conceiving of homosexuality as a form of evil or madness aren't consistent across time or geography should tell us that hostility towards diverse sexual practices says more about our cultural positions than it does about any natural conditions of love and reproduction.

Sinologists have unearthed many examples of love affairs between men, homoerotic stories and poetry, and clearly tolerated acts of male intimacy in Chinese history. While there are examples of anti-homosexual laws appearing at times in China's past, by comparison with the West homosexuality was a valid option for many citizens there. An English statesman named John Barrow wrote of Chinese homosexuality in the eighteenth century: 'The commission of this detestable and unnatural act is attended with so little sense of shame, or feelings of delicacy, that many of the first officers of the state seemed to make no hesitation in publicly avowing it.'

Prior to the Meiji restoration in the late nineteenth century, Japan was similarly progressive in its attitudes towards intimacy between men. When the Jesuit missionary Francis Xavier visited a Zen monastery at Hakata, he discovered their practice of an 'abominable vice against nature' that was 'so popular that they practice it without any feeling of shame'.[34] Future missionaries also expressed shock and dismay at the widespread acceptance of sexual acts between men of all classes and castes, from monks to samurai.

Using the pseudonym Dr Jacobus X, the late-nineteenth-century French anthropologist Louis Jarolliot wrote *The Art of Love in the Colonies*, which described 'a general history of love in the human race' by observing sexual practices across Indochina, the French West Indies, Oceania and Africa. He noted that homosexuality was common in South-East Asian cultures, yet was hardly objective in his writing. On the native cultures of the French protectorate of Annam, which divides Laos from Vietnam, he wrote: 'The Annamite is a pederast because he is lascivious. Here is a sophisticated old race, which was already corrupted. Pederasty is an innate stigma, which the European discovered in full bloom and from which a few (a small number, let us hope), have profited.' Similarly, of the North Africans, he wrote that the Arab was 'an inveterate pederast, even in his own country, where women are not lacking'.

The traditional term *pederast* derives from the Greek word *paiderastia* – literally, 'boy love' – which over the centuries came to refer to any homosexual act. Among the most famous of ancient cultures that tolerated and even embraced sexual diversity were the ancient Greeks.[Δ] Many ancient Greek societies, such as those on Crete and in Sparta, specifically described relationships based on an imbalance of age and power: an older, experienced male coupling with a boy who was coming of age.

Sexual practices and relationships are complex things. They involve economics of love, lust, power and resources. We can try to categorise them, but direct comparisons between borderline (if not outright) paedophilic practices of ancient Greece and modern consensual gay relationships are like comparisons between arranged child marriages and a Catholic wedding. Similarly, the 'insemination' rituals between men and boys of the Papua New Guinean Etoro people hardly bear any resemblance to what we'd accept today as a modern, consensual gay relationship between adults.

---

[Δ] To be more specific, the diverse spectrum of cultures that were spread across the Aegean from the time of Homer (around 600 BCE) to about 400 CE.

Looking to history, it becomes clear that what an individual considers to be an acceptable or unacceptable sexual practice is heavily influenced by their surrounding cultures. Whether it's ancient China, Japan, Melanesia or Greece, or subsets of cultures across time and space, strict adherence to monogamous, adult heterosexuality is far from a universal norm.

So what is the origin of today's widespread vilification of homosexuality? If sexuality is – and has been – so incredibly diverse, why did we start to think of it as an immoral or biological aberration?

One possibility is that sex among social animals like us isn't just about pleasure and reproduction. It's about reinforcing a social order. Think of it as a political exchange, a way to strengthen bonds with those close to us. As such, it's also a firm statement about who we are as a group.

As the glue that binds us, how we have sex can form the basis of sorting 'us' from 'them'. Labelling an enemy as homosexual has been a common form of propaganda within cultures that define themselves as strictly heterosexual. Even that most tolerant of civilisations, the Greeks, debated among themselves the morality and normality of who slept with whom. Xenophon, a student of Socrates, was somewhat critical of the Spartan practice of reinforcing bonds between militants by using pederast pairings.

The Spartan Lycurgus – who supposedly introduced the custom of mentorship among the troops after being influenced by the Cretans – explicitly forbade physical intimacy between the pedagogical 'hearer' and the 'inspirer'. Yet he didn't seem all that surprised by the resulting lack of chastity. 'For in many states the laws are not opposed to the indulgence of these appetites,' he says almost dismissively. Across Greece, as in China, homosexual indulgences were both tolerated and opposed, criticised and romanticised.

Similarly, across the Middle East we can find examples of both condemnation and practice of diverse sexualities. The law of Leviticus (20:13) condemns men who lie with men 'as with a woman', most likely as a clear way to promote Israelite customs over those of neighbouring Canaanite cultures. According to Canadian scholar of LGBT history, Louis Crompton, the word used to refer to homosexual practices in ancient Hebrew scriptures – *kadeshim* – might be translated literally as the 'holy men' of those non-Hebrews over the hill, the Canaanites. Reading between the lines, the connotation was that people like us don't have sex in the way those people have sex.

As Christian teachings permeated the West during the first few centuries CE, sharing the letters of St Paul and various Jewish books became a means of uniting under a single canon, and the sexual distinction of 'us' and 'them' gained strength with the spreading monotheistic identity. The spread of Abrahamic beliefs across Europe and then the world reinforced the belief that homosexuality – among both men and women – was a choice, and one that was best avoided if you wanted to be 'one of us' and avoid punishment.

Not everybody was convinced that sexuality was purely a matter of discretion. The contrasting view – that the kinds of body bits that turn us on are influenced by our biology – dates back further than you might think, with signs of such a debate as far back as ancient Greece.

A collection of problems written in a Q&A format attributed (though somewhat dubiously) to the philosopher Aristotle includes a biological explanation of why some men desire to be anally penetrated: his own semen must be leaking into his anus through pores or passages that are 'not in a natural condition either because those leading to the testicles are blocked … or some other reason'.[35] Another idea was that a man developed a taste for penetration after receiving semen from somebody else. A consequence of this acquired desire, it was thought, was 'effeminate' behaviour: 'for though they are male this part of them has become maimed', the author thought.

Note that the 'problem' being discussed was a desire for anal penetration and the emasculation that came with it. Not homosexual love. Not the act of penetrating, which allowed one to retain masculine characteristics. The primary concern was men becoming too much like women.

A fifth-century-CE Roman physician named Caelius Aurelianus was convinced that the desire to be penetrated was a disease of the mind rather than the body, a result of an improper mixing of the male and female 'seed' on conception. 'People find it hard to believe that effeminate men or pathics really exist,' Aurelianus wrote. He went on to explain how these 'pathics' would attempt to carry on heterosexual relationships, until inevitably, exhausted of their virility, they were driven to give in to their homoerotic urges. Homosexuality has not just been about acts of intimacy or penetration, but about how we present qualities defined as uniquely masculine or feminine.

Aurelianus wasn't a fan of women being overly masculine, either. His book *On Chronic Diseases* also included translations of the work of another physician, Soranus of Ephesus, who recommended a clitoridectomy for oversized clitorises, lest they cause women to take on a more dominant role in sex – or, worse still, reject men altogether.

Skipping forward to the eleventh century, the Persian polymath Avicenna also described the male desire to be anally penetrated as a mental disease, one caused by an early homoerotic experience. He dismissed contemporary physiological explanations that focused on an alleged branching nerve in the penis. 'The men who try to cure these people are foolish,' he wrote, 'for the cause of their illness lies in their imagination; it is not natural.' His cure was simply a regime of punishment to break their lust – sadness, hunger, sleeplessness, imprisonment or even a good flogging would do the trick.

In the fourteenth century, one famous scholar returned to the question of homosexuality in Aristotle's *Problems* and came up with his own hypothesis. While others had written translations and commentaries on this famous ancient work over the centuries, most of them chose to gloss over section 4.26, which starts with the question: 'Why do some men enjoy sexual intercourse when they play an active part and some when they do not?'

But not Pietro d'Abano. His audacious commentary, written in an age when talking about the biology of sexuality was rather taboo, showed the question of sin versus disease was present even back in medieval times. While Avicenna and Caelius had described aberrant sexual behaviour as a diseased 'will' affecting the body's physical nature, Pietro went to great lengths to describe the cause as not just congenital, but anatomical. In his words, the cause of homosexuality was 'not constituted by nature operating in a proper manner'. He cited both Aristotle and the ancient physician Hippocrates to support his argument, stating that 'it does not seem that habit can be converted into nature to the extent that nature would abandon its pristine order'. In other words, Pietro figured that no matter how many times a man was anally penetrated, nature was just too damn strong to change course. It had to be a state of nature to begin with.

While he saw a theoretical cure via surgery, he did warn it wasn't a smart choice, 'for sometimes cutting results in total destruction of the pores, and death'. Not to mention, 'it might happen that the man be turned into a woman, which would be a change for the worse'. Heaven forbid. Instead, Pietro suggested that 'whatever spermatic moisture these men have must be evacuated away' through a mix of diet and desiccating medication.

During his lifetime, the Italian philosopher and physician was accused of liaising with the devil via a bunch of familiars he kept in a crystal container, and of practising astrology and (later) being an atheist. Suggesting homosexuality was a disease of the flesh and not of the mind was just one more item on a long list of grievances the Church held against him. For his crimes against God, poor old Pietro faced trial before the Inquisition not once, but twice. Maybe he was fortunate that he died before the second trial ended. He was found guilty posthumously, and if a friend hadn't hid his corpse it would have been burned. The frustrated Church had to make do with a public proclamation and incinerated an effigy instead.

In the centuries following Pietro's commentary, an increasing number of anatomists, naturalists, philosophers and physicians wrestled with the question of whether 'abnormal' human sexuality was a deviation of the mind or body, making little progress.

By the turn of the twentieth century, the world remained divided on whether homosexuality was a crime deserving severe retribution or a disease deserving torturous treatment. The medicalised terms *heterosexual* and *homosexual* first appeared in 1886, coined by forensic psychiatrist Richard Freiherr von Krafft-Ebing and presented in his textbook on sexual behaviour, *Psychopathia Sexualis*. It's this same text that gave us words such as *sadism* and *masochism* to describe erotic responses to pain. To Krafft-Ebing, any form of intimacy that did not serve the purpose of reproduction was a pathology. 'With opportunity for the natural satisfaction of the sexual instinct, every expression of it that does not correspond with the purpose of nature – i.e., propagation – must be regarded as perverse,' he claimed.

The English physician Havelock Ellis was the first to produce an English textbook on homosexuality (a word he openly disliked). It appeared in 1897 and was titled *Sexual Inversion*; not unlike Pietro d'Abano some five centuries earlier, he took a relatively sympathetic position on homosexuality, based on his belief that it was completely congenital.

The highly influential Austrian psychologist Sigmund Freud disagreed with Ellis, theorising that homosexuality was a developmental deviation of the bisexuality he believed to be inherent in all children. Freud speculated that it may have resulted from a twist on the Oedipal complex, only instead of developing sexual feelings towards one's mother, a son develops them towards members of his own sex.

Even if he did regard it as an abnormality, Freud harboured no judgement towards those who engaged in homosexual practices. What's more, he explicitly denied that it constituted a disease: 'Homosexuality is assuredly no advantage, but it is nothing to be ashamed of, no vice, no degradation, it cannot be classified as an illness; we consider it to be a variation of the sexual function produced by a certain arrest of sexual development.'

In spite of Freud's influence on the developing field of psychology, in the first edition of the *Diagnostic Statistical Manual*, in 1952, physical attraction between members of the same sex was included as a symptom of a psychological illness categorised under the general heading 'paraphilia'. Being gay officially became a disease.

In the radical, post-war, postmodern, counterculture freedom of the 1960s, opposition to this swelled into a rage. In 1970 activists protested the defining of homosexuality as a disease outside the American Psychiatric Association's convention in San Francisco. And again in 1971. By 1973 the tide was shifting, and with the printing of the seventh edition of the *DSM-II*, in 1974, homosexuality was moved into a new category titled 'sexual orientation disturbance'.

Heralded as a landmark event, this change was claimed to be a response to research that cast sexuality in a new light. The truth is a little less dispassionate. The categorisation still permitted psychiatrists to continue to treat homosexual behaviours as an indication of abnormal psychology; they just weren't explicitly labelled as a form of paraphilia alongside paedophilia, fetishism and sadomasochism. The journals of the day continued to print papers on treatments, such as new aversive conditioning procedures, electro-convulsive therapy and promising new pharmaceuticals.

In 1980 another category was proposed. Titled 'ego dystonic homosexuality', it was defined by a persistent lack of heterosexual arousal that interfered with the patient's initiation or maintenance of desired heterosexual relationships, and persistent distress from a sustained pattern of unwanted homosexual arousal.

Only in 1986 were all references to homosexuality removed from the *DSM*, its section rewritten to diagnose persistent and marked distress about any sexual orientation. It might no longer be a disease, but the biology behind our sexuality is still an intense area of research for scientists keen to understand why human sexuality is so diverse.

Recent decades have seen a veritable mountain of studies looking for clear distinctions in the neurology of individuals who identify as gay, straight or somewhere on the spectrum of bisexuality. While there appears to be no single switch-like factor that determines your sexual orientation, there is research that suggests the size of the front part of our hypothalamus, a nub called the third interstitial nucleus, just might gently bias our sexual predilections.[36]

Geneticists, meanwhile, have sought answers in our DNA. A study conducted back in 1993 produced the first hint of a 'gay gene', located on a section of the X chromosome called Xq28.[37] It was a tiny study but drew intense debate. Follow-up studies over the decades have both confirmed[38] and refuted[39] the find, showing just how complex the topic of sexuality and genetics is. Studies on identical twins have suggested there's anywhere between a 7 per cent and nearly 70 per cent chance that if one twin is gay, so is the other.[40] The fact it's not 100 per cent suggests more is going on than just what we inherit from our parents, though twin studies are notoriously prone to all sorts of sampling problems.

In 2015 a team of researchers from the University of California in Los Angeles announced their discovery of epigenetic markers on certain genes – chemical padlocks that can be added in response to environmental influences – that were loosely associated with sexual preferences.[41] The method and the significance of the results have been heavily disputed, and even if such studies have merit, they are far from amounting to discoveries of some elusive 'gay genes'.

Still, taken altogether, studies such as these hint at deep biological influences over the kinds of gender-based characteristics each of us finds arousing, and many are open to being affected by our environment as we develop. No doubt there are numerous factors at work, which vary depending on how the genetic hand we're dealt is affected by the chemical environment in the womb or the social environment after being born. Or, more likely, both. As it's on a spectrum, sexuality isn't going to have some single fundamental cause; rather, it'll be pushed by a vast number of tiny nudges. Some nudges are bigger than others, sure, but they're numerous nonetheless.

It's a stretch to say definitively whether a person is born strictly gay or straight or wavering anywhere in between. But there is little doubt that myriad complexities beyond our control, embedded within our tissues and biochemistry, play a significant role in influencing what gets us off and what gets us soft.

Biologists recognise that sexual behaviour among various species is not exclusively about the act of insemination, but serves purposes of communication, competition and reinforcement of social bonds. Though it's drawing a long bow to compare all forms of male–male or female–female sexual display with intimate, long-term human relationships, the fact that many social animals such as dolphins and chimpanzees can display homosexual pairings suggests nature has something more to say on the matter of sex and gender. As many as a quarter of black swan pairings are between males, for example – often to the extent that they temporarily form a triad with a female swan, only to drive her from the nest once she's laid her eggs. Homosexual behaviour among captured penguins has been of interest to biologists for well over a century.

Studies such as those by sex researcher Alfred Kinsey, who showed in his 1948 report *Sexual Behavior in the Human Male* that homosexuality was common across those of different classes and backgrounds, have gone a long way to reducing prejudice. Today, around 1 per cent of men and 2 per cent of women identify as predominantly gay. A 2016 National Health Statistics Report by the US Centers for Disease Control and Prevention found that 6.2 per cent of men had experienced some form of sexual activity with another male, while 17.4 per cent of women acknowledged having had sexual intimacy with another woman.[42] The report confesses that its figures could be affected by the phrasing of the questions, and vary slightly across cultural and ethnic backgrounds. Counterculture social movements in post-war America have also contributed to reduced stigma concerning diverse sexualities, and more relaxed attitudes towards sex in general. About 60 per cent of the population now views homosexuality as morally acceptable, a figure that is steadily growing.

Homosexuality's shift from sin to disease to normality has been gradual, staggered and patchwork, and remains incomplete. The youth branch of Australia's Victorian Liberals motioned to debate a law change that would allow doctors to 'offer counselling out of same-sex attraction or gender transitioning' in 2018, a move that was swiftly blocked by the state's party president.[43] In 2017 the *Daily Telegraph* in Australia included same-sex attraction in an infographic alongside obesity, drug-use and mental health problems, demonstrating that being gay is still thought of as an illness in many circles, and no doubt will be for some time to come.

Being gay was considered bad when it was thought to be solely the product of a free mind that opted to act against the rules established by moral or religious doctrines. As we came to appreciate the relationship between our biology and our emotions, our appetites and our drives, it became a madness that led us to contravene what was assumed to be natural – a disease that needed fixing.

Far from being a simple binary, though, it's clear that diverse sexual interests and practices are the result of a mix of physical influences and cultural norms. Being gay, bi, straight or even asexual is the product of both an internal system of biochemistry and an external environment of human behaviour. Whether or not that internal biochemistry constitutes a disease depends not on nature but on what we humans collectively value.

# ASHAMED

Pluto is currently not a planet. It hasn't been one ever since the International Astronomical Union declared a planet to be an object that:

1   Was in orbit around a star;
2   Was massive enough to squeeze itself into a fairly round shape thanks to its own gravity; and
3   Had cleared a path around its orbit.

In spite of the sadness of children everywhere, Pluto wasn't bothered. It was the same old ball of rock and ice zipping around the Sun in its weird, wide loop. Unfortunately, astronomers had found other balls of rock and ice bigger than Pluto, forcing them to confront a dilemma: should we add them all to a special list (forcing our children to learn them with increasingly lengthy mnemonics), or should we cut back the number of planets in our solar system to just eight?

In order to understand our natural world, we generally shape our definitions to create unique boxes, and then make rules about how they compare. It's often as messy as it is useful. There are things that are very nearly planets, just as there are things that are borderline stars, comets and asteroids. There are states of matter that are sort of gases, close to plasmas and not-really liquids.

We work together to construct categories for the multitude of objects and phenomena we experience as a species. It's easy to use the term *socially constructed* rather flippantly; diseases are indeed socially constructed concepts, but so too are planets, bugs, rocks, cells, atoms and the colours of the rainbow. It doesn't tell us much to admit that most of the definitions we use in science are invented by groups of people who are heavily influenced by their values and preconceptions.

What's important is that we can objectively describe the path and shape of a planet, or the wavelength of a colour, or the genetics of an insect, and use that to debate their categories. We can see how round an object is out in space, note where it goes and measure the amount of rock and dust nearby. All that remains is coming to an agreement over whether the path is clear enough and the object round enough for it to be officially declared a planet.

To say whether a condition is a disease or a mere variation, we are forced to consider not only its physical nature but also the nature of our own agency and the condition's relationship to some agreed-upon definition of morality.

My wife, to whom I have been blissfully married since 2007, was diagnosed as an adult with attention deficit disorder (a.k.a. ADD, or the mouthful of words it currently goes by, 'attention deficit hyperactivity disorder – predominantly inattentive'). It's a condition more commonly associated with rambunctious ten-year-old boys who won't pay attention in class than professional, middle-aged women. But along with 5 to 7 per cent of kids and adolescents, roughly 3 per cent of adults today are diagnosed with the condition.[44]

Of course, my wife didn't just wake up one summer morning feeling a little more distracted than usual. For most of her life, in fact, she claims to have found concentrating on many tasks a challenge. While she can focus (to an inhuman degree, I might add) on actions she enjoys, keeping her mind on tedious chores is hard, as thoughts flit in and out of her field of awareness like bees in a field of flowers. I've known about her beautiful, whimsical brain since day one of our relationship – often painfully, as I've closed pantry doors she's randomly opened, turned off light switches she's left on, picked up parcels she's forgotten, tried calling the phone still sitting on her bedside table, and encouraged her to persist with studies and job applications as more exciting hobbies have captured her attention.

On a hunch at the ripe old age of no-longer-in-her-twenties, she decided to seek an expert opinion, which led to a diagnosis and consequently a selection of options to help her manage tasks she finds particularly difficult. Crucially, learning how her brain worked didn't reduce the responsibility she felt she had over her actions. In fact, quite the opposite – it allowed her to take better control of them.

Not that I no longer have to close cupboard doors or triple-check if she's remembered her keys, but with a mix of forced habits, occasional use of medication and renewed self-esteem, my partner now has confidence that her attention-deficit brain can deal with most of the curve-balls life throws. She is today a secondary school teacher who manages her brain brilliantly because she understands better how it works – plus she's more sympathetic to students who also struggle to maintain their attention.

So you'd think a person writing a book like this would be understanding towards her neurological condition and the impact it has had on her life, wouldn't you? Umm … if only it were that easy! It's with no small amount of shame that I admit I still have to choke down the occasional sense that her appeal to a diagnosis of ADD is merely a convenient excuse for laziness. After all, my brain has no problem focusing on even the worst jobs. For its many other faults – including being prone to bouts of depression and a strange inability to recognise faces – I have zero problem turning off the light as I leave the room, closing the bin, putting away my dirty socks, checking messages or remembering to pick up the milk on the way home from work.

It takes more than a small amount of effort for me to remember that the practices I find easy to manage aren't necessarily the result of challenging choices I've faced but decisions made easier by opportunity, experience, upbringing and genetics. Even those shaped by experience were opportunities I rarely had much say in, such as my upbringing or chance misfortunes.

In the end, I simply can't say where my brain stops and my own unfettered decisions begin. Nor can I make that call on my wife's behalf. I don't know where the disease of ADD starts and the choice to concentrate begins. I also can't tell you precisely where my ability to abstain from one more beer stops, and where another's compulsion to drink at three in the afternoon commences. Nor where another's control over temper fizzles out and their destiny as a violent criminal starts.

Yet it's hard to avoid using language of mental health when trying to make sense of the world. We talk of unfathomable acts of death and destruction as the work of madmen and lunatics; acts of terror committed by otherwise sympathetic offenders are blamed on mental health problems without any hint of a diagnosis. The United States struggles with the question of whether mass shootings are the result of a breakdown in mental health services or an inevitable consequence of the freedom to own guns. When we conclude that they're a product of madness, we no longer need to ask how a mind could commit such evil; instead, we see a soulless act of Mother Nature, something more akin to a tornado or tsunami than a decision made by somebody's child, parent or sibling.

With a choice between only shame and disease, we're effectively deciding who has ultimate control over suffering. The conversation is less medical and more moral. Rather than leading us to ask how we can work together to ease the painful consequences of our biology, mental illness makes us ask how we should respond to grievances – by exacting retribution against the offender, or by absolving them of their responsibility for their own body?

ADD does make my wife's life harder. And, because she's my life partner, it makes my life a little more complicated on occasion. I hesitate to use the word *suffering*, given the magnitude of pain that poor mental health is capable of inflicting, but we have our moments of frustration. It's not like my wife wants these challenges to impede our way of life, so she makes an effort to limit the impact of her fuzzy brain, changing her habits where she can and taking medications where she can't.

For my part, treating her flavour of 'insanity' as a moral problem is a path to failure, devoid of practical solutions. By the same token, if we say it's a disease that means she can absolve herself of responsibility, leaving me to accept lost keys and tardiness as part of life with somebody with ADD. But a person can still be responsible for mitigating the negative consequences of their brain while knowing there are elements of it beyond their control.

We can be an asshole *and* have Asperger's; be insane *and* criminal, be lazy *and* have ADD; have low energy *and* have clinical depression. We all have a responsibility for shaping how our biology impacts others around us, and for helping others do the same. Similarly, we all have the opportunity to shape how society can accommodate our diverse physiologies, and can encourage others to do likewise.

# IV
# ABNORMAL

# IV
# ABNORMAL

# DIRTY

A lot of people in this world don't care for foreskins. Put every penis on Earth into a room and then pick one out at random, and you'll have a one-in-three chance the wrinkled tube of flesh covering the tip will have been cut off, most likely during its owner's childhood.[1]

Travel somewhere like Israel and you'll have little chance of finding a foreskin, thanks to its Jewish population. On the other hand, try Denmark and you'll be surrounded by them, with something like 97 per cent of the male population uncircumcised. In Australia, New Zealand and Canada, roughly a quarter to a third are methodically cut.[2]

The United States is a mixed bag: from less than one in five to more than nine in ten newborns are circumcised, depending on the state. But averaged out, around 70 per cent of young boys get circumcised, making the country one of the most foreskin-free places in the English-speaking world. It's hard to tell if the practice is dropping in the land of the free, with different studies over the years hinting at fluctuating trends. One thing is for certain: by no means is circumcision on the verge of vanishing. In 2011 circumcision counted among the most common surgical procedures in US hospitals.[3] It's fair to guess that roughly half of all male infants will be circumcised within days of being born.

If you ask most historians, the origin of removing one's foreskin is hidden somewhere in the deep past. We're fairly sure the priests and nobility of ancient Egypt engaged in ritual circumcision 4000 years ago, and there's evidence that a number of communities across the African, Asian and Australian continents have traditionally engaged in various forms of foreskin cutting and removal. It's likely the practice arose in more than one population, and came and went at different times.

But why did it start at all? There's no shortage of hypotheses – and doubtless there's more than one contributing factor.

One thing to consider is that trauma unites members of a community. It also establishes fitness – and therefore worthiness – of those coming of age. Blood and pain inflicted on a defenceless minor could be a way to say to the parents: 'This kid now belongs to all of us – look how strong he is.' Initiation rituals often involve cutting, bloodletting and extreme discomfort. Scarification, tattooing, tooth-chipping, flesh-stretching, binding, piercing, poisoning … people all over the globe, throughout history, have found ways to damage the body just enough to strengthen bonds of empathy through a shared traumatic experience in a show of strength, beauty or fidelity.

It can just as easily go the other way: mutilating captives, whether by castration, branding or the removal of hands, has long been an act of subjugation and humiliation. By disfiguring a person's body in a small way, it's possible they are shamed without being overtly damaged. Could some traditions be holdovers from a time of slavery, a way to remind members of the community of where they came from?

Some arguments are weaker than others. Claiming that circumcision arose for reasons of hygiene, perhaps to avoid the irritation of stray sand or dirt particles, might sound reasonable but is unsupported by evidence. For past cultures, an intact foreskin probably wouldn't have led to any noticeable increases in infection or discomfort, especially when compared with the risks involved in removing it.

That's not to say circumcision could not come to be associated with cleanliness over time, however. Quite the opposite, in fact. Initiation into a religious community is still a key influence for many parents, but others are driven by a gut-sense that foreskins are dirty parts of the body that increase the risk of catching a disease.

A small study conducted in Canada, for instance, found that just under two-thirds of parents claimed hygiene influenced their decision to circumcise their child. Around half of the surveyed parents said preventing infection or cancer prompted their decision, while 41 per cent cited the fact the child's father was circumcised.[4]

In an Australian survey, 96 per cent ranked hygiene as the most significant reason for circumcising.[5] Family tradition played a role for around three out of five parents, and 14 per cent thought it just looked nicer. In the United States, roughly two-thirds of parents circumcise their infant sons for hygiene reasons, with just 12 per cent prioritising a religious influence.[6] Time and again we see hygiene and disease prevention as top reasons for circumcising penises.[7]

If having a foreskin significantly increases the chance a child will suffer a major illness, it'd be easy to see why hygiene would be a rational choice for a parent to make on their newborn's behalf. After all, we advocate the jabbing of needles into babies without waiting for junior to offer an opinion. Unfortunately, the science isn't quite so clear-cut on the health benefits of circumcision. As a parent, it's hard to know where to turn for information on the subject, given that some medical bodies say 'trim' and others equate the practice with mutilation.

In 2015 the US Centers for Disease Control and Prevention (CDC) officially endorsed circumcision as a way to reduce the risk of catching and transferring the human immunodeficiency virus (HIV) and other sexually transmitted infections, and of lowering the chances of developing penile cancer.[8] Along with a set of guidelines for medical practitioners, the CDC promoted the benefits of circumcising infants, which they believed far outweighed the costs, and recommended that doctors tell their uncircumcised male patients in high-risk categories to get the chop.

They based their decision on a handful of studies, the most significant being the substantial amount of research conducted on communities in Kenya, Uganda and South Africa, which concluded that circumcision reduced the transfer of HIV in vaginal intercourse by about half, from 2.5 per cent to about 1.2 per cent. Not everybody agrees that this research itself is sound or definitive, however, nor that it applies so easily to a post-industrial nation that has a comparatively low incidence of HIV. Even the CDC concedes that North American and African countries have significant differences. There are also concerns that overconfidence in the small protective benefits of circumcision might be leading to increased risk-taking and shunning of condoms among African men, further complicating the matter.

What about cancer? While it's incredibly rare, striking less than one in every 100,000 penis owners each year in the United States, and just over one in every 100,000 in Europe, cancer of the penis is not a trivial disease.[9] The risks are raised by having HIV, genital warts, human papillomavirus or inflammation caused by an infection or injury. According to systematic reviews of the literature, circumcision does appear to offer some small amount of protection – the American Cancer Society calculates that for every 900 circumcisions in the United States, one case of penile cancer can be prevented.[10] There have also been hints that being circumcised before becoming sexually active might decrease a man's chances of getting prostate cancer later in life by around 18 per cent.[11]

So the big question: does being circumcised make you in any way safer? The data alone suggests yes ... sort of ... a little bit ... maybe. The problem is that 'safer' needs to take into account not just the benefits but the potential costs. Not to mention less risky ways to achieve the same level of protection, including practising safe sex, responding quickly to infections and maintaining basic hygiene habits.

Critics of the official CDC endorsement argue that it's impossible to know if the benefits outweigh the risks of preventative surgery when those risks haven't been adequately explored. They claim it's inconceivable that the experts behind the advice 'could have objectively concluded that the benefits of the procedure outweigh the risks when the "true incidence of complications" isn't known'.[12]

Finding good data on the actual rates of complications from circumcisions conducted in hospitals is a bit of a minefield; and forget about including procedures performed outside of a medical setting. One study found that about 0.5 per cent of the 1.4 million circumcisions performed in US hospitals in the first decade of the twenty-first century resulted in some kind of adverse effect. This could be as severe as a partial amputation of the penis or as minor as needing to suture an artery.[13] While the source of the data prevented the researchers from determining whether any procedures had contributed to a death, they did mention a previous study that identified a total of three mortalities between 1954 and 1989 as a direct consequence of circumcision.

Is 7000 accidents over ten years a lot? It's hard to tell, as comparing one set of risks with another is far from straightforward. Is it fair to match a few thousand stitches with a few thousand glans infections that might have been avoided if the foreskin had been removed during childhood? Can a theoretical few dozen people who didn't get HIV or cancer that decade be worth the seventy-one partial amputations of the penis identified by the study? For that matter, how do we compare three deaths over four decades with dozens of prepuce tears each year? Science can crunch the numbers, but comparing which course of action is worse is basically a philosophical exercise.

Around the world, many medical associations don't believe that the risks – no matter how few or trivial – are worth it. If you ask the Royal Australasian College of Physicians, they'll unequivocally tell you to forget about it: 'After reviewing the currently available evidence, the RACP believes that the frequency of diseases modifiable by circumcision, the level of protection offered by circumcision and the complication rates of circumcision do not warrant routine infant circumcision in Australia and New Zealand.'[14] The Danish Medical Association has advised that circumcision is a choice that should be left to the individual, not the parents.[15] The Canadian Pediatric Society 'does not recommend the routine circumcision of every newborn male'.[16] The Swedish Pediatric Society has gone even further, requesting that the practice be banned.[17]

Reflecting on the statistics at the beginning of the chapter about the factors that influence parents' decision-making, we might reason that folks in the United States – and to a lesser extent those in Australia, New Zealand and Canada – are either more risk-averse when it comes to sexual health and disease than parents in many other nations; are more misinformed on the true magnitude of benefits versus risks; or are influenced more by factors that make a foreskin seem unusually dangerous and unhygienic. Possibly all of the above could be true.

Why are foreskins considered dirty by some cultures and not by others? A recent period in US medical history sheds light on why so many parents in the Western world believe surgical intervention is necessary in order to maintain their child's health and cleanliness.

In fact, we can trace it back to the morning of Wednesday, 9 February 1870, when the highly respected orthopaedic surgeon Lewis Sayre visited a colleague in New York to give some advice on a patient who had travelled all the way from Milwaukee. 'The little fellow has a pair of legs that you would walk miles to see,' wrote his colleague, describing a five-year-old boy whose legs wouldn't fully extend beyond a 45-degree angle.[18]

Sayre's diagnosis was paralysis of the muscles that normally pulled the legs straight – a fortunate conclusion for the poor boy, really, as his colleague had called on the surgeon to cut the tendons in his hamstrings, believing it was those muscles that had tensed up, locking his legs in place. The problem was that Sayre had no idea what might be causing the paralysis. While he was testing the boy's reflexes, however, the family's nurse warned him to be very careful. 'Don't touch his pee-pee,' she said. 'It's very sore.'

Apparently, sore was an understatement. The poor lad was said to have a penis so sensitive that even the brushing of the bedsheets at night caused him 'painful erections'. Taking a closer look, Sayre noticed that 'the glans was very small and pointed, tightly imprisoned in the contracted foreskin, and in its efforts to escape, the *meatus urinarius* had become as puffed out and red as in a case of severe granular urethritis'.

He considered 'excessive venery' – a formal way of saying sexual indulgence – to be a 'fruitful source of physical prostration and nervous exhaustion' that could lead to paralysis. Putting it another way, painful erections could be what was causing the kid's leg paralysis. Instead of cutting the child's tendons, the eminent orthopaedic surgeon would merely trim the skin from the end of his penis. In fact, Sayre was so confident in his solution that he took the boy to the Bellevue Hospital so others could observe his handiwork.

In the following days the young patient's health was reported to have improved, until several weeks later he was walking again. Sayre concluded that the circumcision 'quietened' his nervous system by relieving his imprisoned glans.

Clearly believing he was onto something, Sayre barely waited until his patient was fully upright before trying the procedure on the partially paralysed teenage son of a local lawyer. Since electric shocks and strychnine injections into his muscles had failed to cure him, Sayre figured circumcision was worth a shot. The father was worried that masturbation was the problem; what better way to keep his hands away from his manhood for a few weeks than to slice into it with a scalpel? Whatever the cause was, the adolescent reportedly recovered within weeks, showing such an improvement in health that it was said his most intimate friends scarcely recognised him.

Over the following months, Sayre continued to perform circumcisions in the belief that it was the solution not only to chronic muscular conditions, but to a variety of other illnesses and ailments as well. 'Many of the cases of irritable children, with restless sleep, and bad digestion, which are often attributed to worms, are solely due to the irritation of the nervous system caused by an adherent or constricted prepuce,' he claimed.

Sayre went even further. 'Hernia and inflammation of the bladder can also be produced by the severe straining to pass water in some of these cases of contracted prepuce,' he noted. He published his proposal in *Transactions of the American Medical Association*, hoping to draw the attention of the medical establishment to what he saw as a significant impediment to health and wellbeing.[19] In his later years he continued to add more afflictions to his list, including mental health problems.

With the benefit of hindsight, it's easy for sceptics to dismiss Sayre as a crackpot occupying the fringes of medical science. Yet in the late nineteenth century he was regarded as anything but a quack: he had a prestigious career as a New York surgeon; he became a government official by appointment of the mayor when the Civil War erupted; he ordered reforms in sanitation, smallpox vaccination and quarantine measures; he designed cutting-edge orthopaedic treatments; and he eventually became the president of the American Medical Association. To put a cherry on top, he was largely responsible for turning their publications into what he named the *Journal of the American Medical Association*, which remains a highly respected science journal to this day.

Given the man's reputation as a medical expert, it's easy to see why Sayre's 'discovery' was taken so seriously. But there were several other factors that saw his circumcision cure-all spread not just across the United States but throughout the Western world.

One was practical – surgeons cut because now they could. By the late nineteenth century, germ theory and the age of sterilisation meant surgery had gone from being a desperate act of life preservation to a means of treating less threatening disorders and diseases. What's more, innovations in anaesthesia had made patients more amenable to going under the knife; no longer did they need to be inebriated, held down or desperate. A whiff of chloroform and out they went. Surgery could now be a delicate act of dissection rather than a wrestling match.

Because surgeons could now slice into patients at will, cancerous tumours were removed long before they promised death. Entire organs were pulled out if they were thought to be contributing to some real or imagined condition. The foreskin was just another victim of this eagerness to craft the body in the absence (mostly) of killer infections or severe pain.

The second factor was theoretical. A popular medical hypothesis was a concept called reflex neurosis. In simple terms, it claimed that an energy of some variety flowed between each and every organ in the body by way of the nervous system. It was a reasoned attempt to describe the complexities of nervous impulses, given what little was known about electrochemistry and cytology at the time. Following on from this assumption, each organ, muscle or piece of skin was presumed to affect the health and function of the other organs in the body, far and wide. If you think this sounds familiar, you only need to go as far as your local reflexology clinic or chiropractor to see how the belief has persisted, even after science has long since moved on. In New York back in the 1870s, the nervous affinity model aligned perfectly with Sayre's observations.

Lastly, the spread of medical circumcision was helped by a good dose of old-fashioned moral righteousness. Promiscuity, sexual deviancy, masturbation, licentious desire … there was a line between healthy lust and teenage depravity, and the latter could be nipped in the bud with surgery.

Women in various cultures have suffered surgical interference in the name of sexual 'treatment' far more often than men, with clitoridectomies readily conducted without concern for 'unsexing' the patient. Still to this day, girls are forced to undergo brutal procedures that remove a part of the body seen as dirty and immoral. Men, on the other hand, have only ever been castrated in the most extreme cases of depravity or illness.

Yet circumcision, too, was once viewed as a way to curb excessive sexual desire, especially where self-pleasure was concerned. John Harvey Kellogg recommended circumcising small boys in his 1877 publication *Plain Facts for Old and Young*: 'The operation should be performed by a surgeon without administering an anaesthetic, as the brief pain attending the operation will have a salutary effect upon the mind, especially if it be connected with the idea of punishment, as it may well be in some cases.'[20]

Kellogg promoted sexual abstinence for a healthy lifestyle, going so far as to develop a bland form of nourishment that he hoped would reduce arousal. His famous 'toasted corn flakes' might not have been boring enough to stop your average adolescent from pleasuring himself under the sheets, as he'd hoped, but they did become a well-recognised product on the breakfast table once his brother, Will, ran off with the recipe to build a cereal empire.

All these factors helped convince physicians that Sayre's wisdom was worth taking seriously, so off they went in search of unusual foreskins that might need to be removed. The diagnosis they commonly offered parents was called *phimosis* – a narrowing of the foreskin that made it difficult or impossible to retract behind the glans. While this condition does cause problems following adolescence, today it's considered to be relatively rare – affecting less than 1 per cent of fully developed penises, in fact.[21] In young children, it's now considered normal to be unable to fully retract the foreskin until about age seven, or even until the mid-teens, making 'infant phimosis' an oxymoron.

Yet in late-nineteenth-century America, thanks in no small part to Sayre, most male infants' penises were diagnosed with phimosis if they looked in any way abnormal. Which was to say, most penises deserved to be circumcised. So great was the anxiety, and so low the regard for complications, that circumcision was routinely practised as a preventative measure, phimosis or not. Within decades it had become a standard procedure in hospitals for doctors to circumcise newborn males, and phimosis was regarded as the cause for all manner of urinary tract problems, all because some foreskins didn't have the 'right' shape or size.

Physicians no longer blame foreskins for leg paralysis or deviancy, but the culture became normalised. In fact, the circumcised penis became so common that it wasn't just the average aesthetic, it became associated with a standard of cleanliness and good health. Foreskins in American culture are now commonly associated with germs, smells, bodily fluids and poor hygiene ... all of which can be dealt with by simply doing away with the offending bit of flesh while the child is young. And who can blame folks for wanting the best for their kid?

The decisions we parents make are heavily influenced by what our own parents did for us, and by what our immediate tribe emphasises as important. We're forced to rely on the standards handed to us by our community, and to prioritise what we think is popular, average and traditional. We all want our kids to be normal, and we all know what normal looks like.

Pinning down exactly what is normal in definitive terms is a little harder. On one hand, it's what we identify as numerous. But sometimes normal is the midpoint between extremes. We also regard ideas that have stood the test of time as normal simply because they've been around for so long. Normality gives us safety in numbers, both extant and historical, and security in predictability.

But normality is also an illusion, emphasising a vision of correctness without having much to say on *why* it is good or right. In the United States, parents circumcise their children because – among many other reasons – it's the normal thing. No parent wants their child to look at their peers or their kin and feel different or dirty.

Not all abnormalities are diseases, yet all diseases, disorders and disabilities are by definition deviations in numbers, extremes or novelty. Sometimes the lines can blur, and parents feel compelled to make a choice. Do they want their children to suffer the hostility of a society that exiles them as being so different that they're effectively broken? Or can they 'fix' their kids so they fade into the bell curve of average statistics?

# CUT

In the days following 22 August 1965, the Reimers faced the same decision as millions of parents in modern America. Having moved to Winnipeg, Canada, to put some distance between themselves and their rural Mennonite background, the couple was no longer under religious pressure to circumcise their newborn twin boys, Bruce and Brian. However, seven months after their birth, new mother Janet was concerned that the pair seemed to be distressed when urinating.

The two boys were subsequently diagnosed with phimosis, and in April the following year they were admitted for a routine circumcision. In an unfortunate twist of fate, the physician they were expecting wasn't present, so a general practitioner stepped in to perform the procedure. He did so with an electrocautery needle, which used a current to heat a metal probe and cauterise the wound as it parted the tissue, rather than the standard scalpel.

The first incision into Bruce's foreskin didn't cut, so the doctor raised the current. The consequence sparked one of the most famous case studies in recent medical history: Bruce's penis was so heavily burned down the shaft that he had to be fitted with a catheter. His brother, Brian, in the end didn't undergo the procedure. His urinary tract condition cleared up on its own over the following months. Within several weeks of the incident, the tissues making up Bruce's penis died and fell away.

The first modern attempt to surgically reconstruct a penis from skin, blood vessels and fat had been performed a whole three decades earlier, in 1932, but little progress had been made since then on making a phallus that looked and functioned like the real thing.[22]

The head of the hospital's Department of Neurology and Psychiatry advised Ron and Janet Reimer that their son would never be able to have a normal sexual life or consummate marriage. According to the trusted medical authorities of the time, a boy like Bruce would never be considered normal so long as he didn't have a penis. And that, it seemed, was enough for his family to hide themselves and their child away from the community.

The twins were barely a year old when Ron and Janet saw an interview on a current-affairs program featuring a psychologist by the name of John Money, a New Zealander who conducted research on gender at Johns Hopkins University, in Baltimore. The doctor was a pioneer in the modern concept of gender identity – the notion that a person has a strong sense of being either male or female – and was a recognised authority on the psychological impact of possessing ambiguous genitalia.

The interview topic was gender reassignment surgery, with Money claiming that a person's assigned sex correlated with their sense of gender better than their genes. In other words, our individual sense of being male or female didn't always go hand-in-hand with what was under our clothes. As the current-affairs program progressed, the discussion turned to a condition described as intersex.

Money proposed that regardless of a person's chromosomes, a newborn with genitalia that was neither clearly male nor female could have their gender decided for them. He theorised that children were born gender-neutral and adopted their identity based on a mix of social conditioning and fluctuating hormones. If raised to do girl things and if given the right regime of drugs, a child who was anatomically a boy would see themselves as female and possibly be attracted sexually to men, and vice versa.

Not surprisingly, Money had his critics. For one thing, he was a radical, a cliché of the swinging sixties who openly advocated free love and nudism. The current evidence and popular theories didn't match his perspective, either. While Money was making his case, a young postgraduate also at Johns Hopkins named Milton Diamond was providing evidence that gender-specific behaviours in guinea pigs were determined by hormone levels in the womb, meaning the rodents seemed to be born with their gender behaviours already well and truly established.

It's easy to see why Ron and Janet found Money's hypothesis so appealing. Surgery might not be able to create a working penis for Bruce, but it was feasible to make a passing vulva and outer female genitalia. A future life as a female might be easier for their son than life as a male without a penis. They wrote to Money, who saw in Bruce not just a unique case, but an opportunity to show the upstart Diamond that he was wrong.

Twins are nature's perfect template for an experiment. Brian was Bruce's negative control, an example of the boy Bruce could have been, had nurture steered him in that direction. Two months before his second birthday, Bruce received a bilateral orchidectomy, in which his testes were removed and a vulva was fashioned from the surrounding skin. He was renamed Brenda, and given a dress and dolls to play with. His hair was permitted to grow long. To all appearances, the Reimers now had a daughter.

According to Brian, though, Bruce continued to develop behaviours that society would view as masculine: he preferred the toys his brother played with, he sat and walked in a rather unladylike manner, and he even preferred to urinate facing the toilet. What Money dismissed as Brenda being a tomboy his brother and peers saw – with hindsight – as Bruce being a boy.

Money regarded his experiment a success, writing about it extensively during the 1960s and '70s as the 'John/Joan case'. Just like Sayre, Money's influence, charisma and reputation helped his claims gain attention and global acceptance. The now-famous psychologist's conclusions also resonated with developments in the burgeoning field of behavioural psychology, as second-wave feminism began to see science as an opportunity to argue that differences between men and women were a consequence not of biology but of history.

Textbooks around the globe began to reflect Money's theory that gender was something one could change through conditioning, and doctors declared a generation of intersex newborns to be female, confident that the babies' upbringing would only reinforce what surgery had predetermined.

Diamond remained a fervent and vocal critic of Money's work, however. Decades later he investigated and debunked many of the claims made by Money, arguing that gender wasn't something society could easily shift with dresses and dolls, or by crafting a vulva and vagina from a penis: in his view it was determined by numerous factors that had already left their mark long before the child was born.

Meanwhile, the Reimer twins gradually awakened to the experiment unwittingly imposed upon them. Frequent trips to the hospital were an event their friends never had to deal with, after all. Other children bullied the child they knew as Brenda. Janet later recalled: 'They wouldn't let him use the boys' washroom, or the girls'. He had to go in the back alley.'[23] The siblings remembered interviews with Money as being friendly when they were conducted in their parents' presence, but hostile when not. Disturbingly – although this was denied by Money – there were allegations made that the doctor invited the six-year-old twins to simulate sex with one another after presenting them with sexually explicit images, all with a view to reinforce their gender identities.

With the onset of puberty, following years of problems at school, and moving cities (and back again), 'Brenda' was instructed to begin a course of the hormone oestrogen. He knew he was somehow different, and had been told by his father that this was due to 'some mistake' – an event he judged to be violence enacted on him by his mother at some young age. For Janet, this accusation was the final straw. The family stopped seeing Money, who eventually ceased discussing the case in the 1980s. Sadly, for the twins, especially the child born as Bruce, the damage was already done.

Now in his early teens, Bruce was finally told the truth about that 'mistake'. Ron took his son out for an ice-cream, and explained it all through his tears. Bruce sat quietly, staring, feeling every emotion from anger to relief. He finally asked what his name had been before he was called Brenda. Immediately he began presenting himself as the gender he'd been assigned at birth, adopting a new name in the process: David, after the Biblical boy who vanquished a giant.

David Reimer's story doesn't have a happy ending. On learning of the fame of his case study in 1990, David stepped into the public spotlight to share his side. Although Dr Money was invited to meet David, he refused to see the patient who had earned him his fame. In 2002 David's brother, Brian, passed away, most likely from a brain haemorrhage, although there were rumours of suicide. David shouldered the blame, and after two years of spiralling into depression, he took his own life on 4 May 2004, aged just thirty-eight.

David wasn't born with a disease in any officially recognised sense. He wasn't dealt a selection of genes that gave him immense discomfort, nor did he suffer the effects of low oxygen, a virus or a toxin. In every respect, the baby known as Bruce was what physicians would call normal. The tragedy lay not in random events that caused him to suffer, but rather choices made by people on his behalf in response to what they saw as differences they could not tolerate: his penis was abnormal enough to risk removal of its foreskin; his absence of a penis was abnormal enough to decide that he couldn't be a boy; his identity as a boy was abnormal enough to mean that he couldn't be accepted as a girl.

Gender is a hot political topic in today's world, as society polarises over whether transgender women can be trusted to use female amenities, or possess a true feminist perspective as a result of their 'male' upbringing. Half a century on from John Money's theories, there is still a tug-of-war over whether gender is fundamentally determined by nature or shaped by nurture – whether it is a binary category or if there's a spectrum that demands recognition.

Clouding the issue are the diverse contexts we employ gender to describe, not just in our culture but across all cultures. Language struggles to cover the full range of experiences and observations across all facets of society, and we adopt prefixes such as *cis* and *trans* to attempt to distinguish between gender as assigned by others and gender as we personally identify.

If we stick to a strict biology-textbook mode of speaking, the adjective *male* typically labels a variety of related physical characteristics in humans: possession of a Y chromosome; production of sperm; presence of a substantial phallus; relatively elevated concentrations of androgen hormones such as testosterone; and secondary sexual characteristics associated with elevated testosterone levels, such as body hair or an inverted triangle body shape. Behaviourally, the word *male* is associated with a desire for sex, with postured displays of strength, with resistance to pain, and with dominance and aggression.

On the other hand, the adjective *female* labels paired X chromosomes; production of ova; possession of organs such as a womb and ovaries; absence of a phallus; fluctuating concentrations of the hormones progesterone and oestrogen; relatively large amounts of breast tissue; and an hourglass-shaped physique. Functionally, being female is associated with submission, nurturing and a lack of sexual desire.

These generalisations are regarded as fundamentally connected: the tiny Y chromosome is home to just a handful of genes, including one that produces a protein called testis-determining factor (TDF), more commonly described today as the sex-determining region Y (SRY). Linked up with a couple of other proteins, TDF/SRY regulates something called anti-Müllerian hormone, preventing the embryo's undeveloped sex organs from turning into ovaries, changing them into testes instead.

Testes pump out higher levels of the hormone testosterone, which another structure called a tubercle converts into dihydro-testosterone. This ramped-up version of the hormone turns the tubercle into a penis. Without it, the organ grows into a clitoris. Testosterone also clicks into a receptor on many cells, causing cascades of processes that can add muscle tissue, affect brain development and push out coarse hair.

In other words, by expressing SRY, a body can develop anatomical features we label as male. Without an SRY gene, those structures grow into features regarded as female. Having a functioning form of the gene could therefore be considered relatively binary – you either have it or you don't. And if we're generalising by what's visible, then we might be tempted to assume that most humans with a Y chromosome present physical characteristics regarded as male, while having a pair of X chromosomes normally leads to having body bits broadly considered female.

Variety being the spice of life, however, *normally* isn't the same as *universally*. We aren't all textbook examples. And while there is contention over the exact fractions making up such variation, there is no argument that it exists.

As many as one out of every 20,000 owners of a Y chromosome also has a mutated version of a gene for the receptor that would usually recognise testosterone, producing what's called androgen insensitivity syndrome.[24] Cells with this version of the receptor won't detect the hormone, meaning that even if you've got a Y chromosome containing a working SRY gene (that is, what we might term a 'normal' chromosome), chances are your body's tissues won't be affected by the boosted levels of testosterone, and you'll instead develop physical characteristics regarded as female.

You don't need to have androgen insensitivity syndrome to respond to testosterone in unique ways, either. Genetics being what they are, hormone levels can vary radically for a whole variety of reasons, not to mention how different bodies respond to them. An increase in oestrogens coupled with a decrease in androgens in males can produce a condition called gynaecomastia, which is commonly described as 'man boobs', just to emphasise the normality of breasts being things only women have. Conversely, in women a condition of excessive hair growth, hypertrichosis, once led to bearded ladies being featured in the travelling sideshows of old – again, this emphasises that facial hair is something only men tend to grow.

Thanks largely to this impression of hair as abnormal, women in many cultures around the world and through the ages have removed hair from the lower half of their faces, their armpits, their legs and their pubic area. A survey conducted in the United Kingdom found that nearly all women had removed at least some hair during their life: about 90 per cent reported having removed hair from their legs and armpits, and a little over 80 per cent removed it from their pubic area and eyebrows, statistics that are reflected by populations in the United States and Australia.[25] One US survey claimed that nearly two-thirds of women trimmed, plucked, waxed or shaved off all their pubic hair.[26]

Why? The reasons varied from a belief that their partner would prefer it (21 per cent) to personal aesthetics (32.5 per cent). But the biggest motivation, at 62 per cent, was once again hygiene. Having hair 'down there', much like the foreskin, was thought to be just plain dirty.

Not all cultures feel this way. Many variably promote hair-removal as a choice that need not be reinforced by community expectations. Among Sikh women, removing the hair your creator put on your face is seen as an offence, leading many to display anything from a light fuzz to a full beard.

There are cases where the full impact of the SRY gene isn't observed until adolescence. In the south-west of the Dominican Republic, there is a community where being born with a vulva and not a penis doesn't automatically make you female. Here, the words *guevedoces*, translated as 'eggs [testicles] at twelve', and *machihembras*, or 'first a woman, then a man', are used to describe individuals with XY chromosomes who appear female at birth, only to develop more masculine features at puberty.

This development is the result of a non-functioning form of an enzyme called 5α-reductase, which in a working state would turn testosterone into dihydro-testosterone. As the body develops in the womb, the testes still produce testosterone. But the tubercle doesn't get that boost it needs to transform into a penis, allowing the anatomy to take the default pathway to become a labia and clitoris instead. It's only later, on hitting puberty, that a second surge of testosterone enlarges the clitoris so that it resembles a small penis, and causes the testes to descend, the voice to deepen and the body to grow coarser hair.

These Dominicans are raised to fit the model of a young woman, only with the caveat that one day they could take on a more masculine cultural role – wearing men's clothing and doing men's work – as their body changes. Much of what anthropologists know about this community is a result of the work of Cornell University professor of medicine, Dr Julianne Imperato-McGinley, who interpreted her findings as evidence of gender identity remaining flexible until teenage years.[27]

Not everybody sides with Imperato-McGinley's conclusions, though. Gilbert Herdt, an American professor of human sexuality studies and anthropology, has long pointed out that the realities of the *guevodoce*'s life are a little more complex.[28] The fact that these children might be relatively common in this part of the Dominican Republic doesn't necessarily make their passage into adulthood a smooth ride. He reported that out of eighteen subjects studied by Imperato-McGinley, one still identified as female after puberty.[29] Another continued to wear women's clothing, while a third ostracised himself from the community.

A people in the highlands of Papua New Guinea also share the 5α-reductase mutation, thanks to a reduced gene pool; in Melanesian Pidgin the children are called *turnim-man* (which derived from the English 'turn him man'). There, in contrast to the *guevodoce* children, the status of at least the handful of individuals observed by anthropologists is heavily stigmatised, with men who develop penises later given a lower status, and unsubstantiated claims that children born with genital abnormalities are routinely killed at birth.

Throughout history people with diverse physical sexual anatomy have been treated as if they're either cursed or blessed, as a source of fascination, fear or fortune, as monsters or divinities. On rare occasions they are even accorded a distinct gender.

Many Native American cultures have classified gender roles in ways different to the anatomically prescriptive male/female model of Europeans. To the Navajo, a matrilineal culture, people could physically be male, female or an intersex they called *nádleeh*, a word that means 'to constantly transform'. *Nádleeh* themselves are considered a diverse group, divided into those who are predominantly male or female, with rules for sexual pairing based on this distinction. Homosexuality has been regarded as taboo in Navajo tradition, yet *nádleeh* can have sex with both men and women, if not other *nádleeh*.

Examples abound across the world of gender models that don't fit Western ideas of masculine or feminine. In India, the *hijra* are traditionally regarded as those born as intersex males who undergo castration at a young age. In Samoa, a gender category who identify as *fa'afafine*, meaning 'in the manner of women', are recognised and respected. *Fa'afafine* are born male but show a preference throughout childhood for tasks and play culturally seen as feminine; they often later identify as possessing a mix of masculine and feminine traits.

Herein lies the challenge for modern researchers: are these contrasts between anatomy and gender identity the result of social and hormonal forces in early childhood, as Money thinks? Is it significantly the product of chemical influences on a developing foetus, as Diamond claims? Or is it malleable, affected by culture throughout childhood into adolescence, as suspected by Imperato-McGinley?

The science here – as usual – can be rather opaque to us non-experts, with some fields still hotly contested. Women might, on average, have a larger hippocampus – a kind of control station for signals criss-crossing the brain.[30] Or it might be the opposite.[31] Then again, perhaps there's absolutely no difference at all.[32] Research remains fairly divided on whether there are meaningful patterns in the data on some brain differences.

On the other hand, in some fields there's a degree of consensus. Even accounting for their larger body mass, men tend to have slightly more brain volume.[33] We also have a slightly thinner cerebral cortex.[34] There seem to be differences in the production and use of the neurotransmitter serotonin, which governs a wide variety of functions, including moderating sleep and mood, and playing a role in learning.[35] We can be fairly sure there are a bunch of differences in blood flow to some areas, connections in others, density of neurons in specific areas ...

For the most part, there are just two things we should take from the scientific discussion on gender, grey matter and genetics. Firstly, the haze of statistics based on neurological measurements and gender identities isn't easy to sort into simple categories. The literature is rife with debates over where to draw lines, which models to use and how to tease out our biases. If the divide between male and female brains was crisp and self-evident, the discussion would have been settled long ago.

The more important take-home message is that we shouldn't confuse that for an absence of difference. Somewhere between the push of our genes and the pull of society, the physical features of our brains get warped and wired into configurations that can be generalised into some categories relating to gender. Our brains aren't all the same – there are patterns amid the noise, and frankly there isn't a great deal any of us can do about it.

Does that mean our neurological differences can account for an individual's uncomfortable sense that their body doesn't easily match the map in their mind? Was David born with a brain that fitted masculine functions, no matter what his body was shaped to do? Is it possible that an embryo with an SRY gene can grow a brain that doesn't suit the body it moulds, and vice versa?

Some investigations indicate there are variations particular to transgender individuals. One study on eighteen untreated transgender men found the pattern of white matter – the supportive structure in the brain – more closely matched that of twenty-four cis-gender men controls than it did nineteen cis-gender women.[36] A similar study using brain scanning technology on twenty-four untreated transgender men, eighteen transgender women, twenty-nine cis-gender men and twenty-three cis-gender women also found matches in the thicknesses of their cerebral cortices.[37]

It'd be a mistake to oversimplify this as 'male brains in female bodies', given the rich mix of neurological characteristics. Instead, transgender men and women appear to have brains with a variety of characteristics associated – again, statistically speaking – with what a broader society reduces to a split of male and female traits.

The question remains: when and how do these diverse neurological characteristics develop? Are they set in stone in our genes, or are they malleable by hormones and experience? Can we influence our fate, or is it sealed the moment we're conceived?

Unfortunately, studies on transgender genetics are few and far between. In 2013 Diamond did an analysis on 113 sets of identical twins, and found that 33 per cent of those who were assigned a male gender at birth but who later identified as female had a twin who was also transgender. For those assigned female at birth, this figure was just under 23 per cent. While it winks slyly at the possibility of a genetic role, it doesn't completely rule out social and environmental factors in early development.[38]

An earlier study on 112 transgender and 258 cis-gender men found that those who were assigned female at birth were a little more likely to have a slightly different gene responsible for making the testosterone receptor. Again, no smoking gun for a 'transgender gene', but a nod at a possible influence that could lay the groundwork before a person is born.[39]

Conceiving of gender as strictly binary means the term struggles to describe an overwhelming number of concepts, of which only a handful are biological, and only one of which is a strict dichotomy: the possession or absence of a functional SRY gene. It also covers the typical effect of that gene in creating a penis or a clitoris prior to birth. Gender also describes the extent to which testosterone has influenced the production of other hormones, the growth of body hair, physique, strength and brain development.

Then there is the knock-on effect of those characteristics on the behaviours and cultural roles across diverse populations. For most of us with a Y chromosome – even if we have man-boobs and soft fingers, even if we love babies, even if we cry at the end of *E.T. The Extra-Terrestrial* – our brains 'feel' masculine. By the same token, in spite of sometimes having fuzzy upper lips, broad shoulders, a hatred of flowers and a love of football, most people with two X chromosomes experience no great anguish over thinking of themselves as female.

Somewhere between 0.05 and 0.5 per cent of people across the globe experience discomfort thanks to the friction caused by how their mind sees their gender, and by the image of gender imposed by flesh and culture.

The World Health Organization categorises this conflict – which is called 'gender dysphoria' in the *DSM-V* – as a disorder called 'gender incongruence' in the new 'sexual health' chapter of its current compendium of diseases, the *ICD-11*. This is a marked change from previous editions, where it had been categorised under mental health conditions. This isn't the first leap. In 1980 it was moved from 'sexual deviation' to 'psychosexual disorder', then to 'sexual and gender identity disorder' in 1994, and to 'gender dysphoria' in 2013. For now, the tradition of describing at least some aspect of the transgender mind by reference to mental illness persists.

Commonly throughout the ages, being a transgender person has been synonymous with homosexuality, sharing its shame and sinfulness over emasculation, followed by pity and impotence as it became a disease. As such, it has also been more associated with those born male than those born female.

In many cultures past and present, possessing masculine qualities has been natural and something to aspire to. So women who dress in men's clothing or adopt male characteristics haven't typically been regarded as transgender men – they have merely acted in accordance with an established order, even if it was sometimes regarded as a little unusual.

Freud theorised that young women developed anxiety when they realised they didn't have a penis – a stage that was normal, even if transitory. A famous nineteenth-century Irish surgeon by the name of James Barry was born Margaret Ann Bulkley; although this was technically a secret during Barry's lifetime, many later claimed they knew about it. Women could suffer from hysteria, neuroses and ambitions of grandeur, yet only men could suffer from sexual perversion or emasculation. A variety of other factors continued to affirm this bias: throughout the twentieth century, surgery has favoured transforming male anatomy into a female form; most academics in the field have been male; social conventions have made it easier for those perceived as women to present a masculine identity; and transgender men have needed to express a greater degree of distress for there to be considered a need for medical assistance.

The name Magnus Hirschfeld stands out in history as a turning point in transgender advocacy. In 1897 the German-born Jew founded the *Wissenschaftlich-humanitäres Komitee* ('Scientific Humanitarian Committee') in Berlin, in order to campaign against the legal persecution of transgender and homosexual citizens. Hirschfeld saw homosexual men such as himself as effeminate by nature, but deserving of understanding and sympathy rather than persecution. Not that everybody felt that way, with one individual splitting off to form the *Bund für männliche Kultur* ('Union for Male Culture'), based on the principle that being gay was more manly than being straight.

Nonetheless, throughout the ages, men who acted or dressed in a feminine manner have been regarded as mentally defective. Think of good old Corporal Maxwell Klinger at the 4077th MASH in the popular long-running television series: for seven seasons the character used cross-dressing as a way to demonstrate he was mentally unfit for service, a trope that served as a convenient symbol for insanity as well as providing comedic relief.

For a condition to be considered a disease or disorder in the *ICD* and *DSM*, there is an argument that it needs to be an unsolicited abnormality that produces such discomfort that it interferes with the individual's ability to meet society's expectations. In other words, there needs to be something behind the 'dysphoria' part of the diagnosis 'gender dysphoria'.

There's no argument that transgender men and women can experience horrific amounts of distress throughout their lives. A study conducted on 250 transgender women attending the

Condesa Specialized Clinic in Mexico City found that 83 per cent experienced relatively high amounts of distress, with 90 per cent reporting some kind of family, social or work dysfunction.[40]

However, one of the report's researchers, Geoffrey Reed, found that by applying mathematical models they could predict which subjects suffered and which didn't. Significantly, being a transgender man or woman wasn't the determining factor in who experienced distress; other factors were at work. 'We found distress and dysfunction were very powerfully predicted by the experiences of social rejection or violence that people had,' Reed stated. 'But they were not actually predicted by gender incongruence itself.'[41]

As with being gay or bi, being transgender doesn't necessarily lead to physical distress or dysfunction. Unlike ongoing conditions that result in severe pain, rob mobility or inhibit our senses, a significant proportion of the 'dysphoria' component of being transgender is a direct result of prejudice within the greater community.

That isn't to say the incongruence between body and mind would be of no concern in a perfectly accepting society, of course. Nor is simply wiping gender dysphoria from the official books on disease a straightforward solution. It's not even a positive one. For all of the stigma a transgender identity attracts as a listed condition, it is currently the way we justify the medical assistance provided by the medical community.

Shoehorning our bodies into a binary of broken versus normal forces us to claim we have a deficiency in order to receive assistance. The spotlight falls on individuals as impeded and in need of repair, distracting us from the more important question of whether our society's expectations and values might benefit from a rethink.

And that is a problem.

# FAT

Kritios Boy isn't much to look at. He stands just under a metre in height, is made of marble, has lost both his arms and feet, and at some point had his head detached from his body. But he represents a subtle transition in fashion among ancient Greek artists: the moment the modelled human form went from uncomfortably rigid to loose and lifelike.

Around the time of his creation in the fifth century BCE, something changed in Greek culture that prompted artists to sculpt bodies that almost seemed to breathe. Lips bent into slight smirks became more relaxed; muscles and ribs virtually drew in lungfuls of air under a thin covering of skin. But the biggest change was that the weight of the body seemed ever so slightly to fall on one leg, curving the hips and spine in a form sculptors called *contrapposto*.

Statues in antiquity were studies in human perfection. The subjects were typically intended to be beautiful, divinely so, representing ideals of human anatomy. Kritios Boy represents the beginning of an art movement that did more than capture the human form – they captured *movement*, bodies poised with such delicate imbalance that you couldn't help but expect them to turn their head, take a step or throw a javelin.

Oddly, for all the pain artists took to depict life in their creations, many were deliberately unrealistic. Breasts and genitals were undersized, in displays of sexual moderation. Some muscles were removed. Others were impossibly defined. Grooves ran deep to enhance the appearance of strength, even as the sculptor ignored parts of the skeleton, such as the tailbone, or built muscle groups no athlete could ever hope to enhance. Proportions ignored biology in favour of more elegant mathematical ratios, stretching thigh bones and torsos to impossible lengths.

Today we take for granted that art isn't a faithful impression of the world, especially with airbrushed photographs or special effects enhancing physiques and actions in films and on magazine covers, and yet our most popular forms of art still set a standard for normality that nobody will ever attain. For anthropologists and historians, however, examples of art through time and across cultures can describe which ideals in beauty, health and strength are conserved, and which change with our experiences and values.

Take, for example, that most celebrated of human forms in the West throughout the twentieth century, the comic-book superhero. The era of the Great Depression and World War II witnessed a 'golden age' of the comic book, giving birth to perennial favourites such as Superman and Batman, not to mention the Phantom, Captain Marvel and Wonder Woman. Flick through the pages of these early comics, however, and you'll see very different figures to those that were drawn later in the 'silver' and 'bronze' eras of the 1950s, 1970s and today. Muscles weren't crisply defined, with Superman looking almost pudgy, contrasting with the vein-streaked, six-pack-studded heroes and heroines of the modern X-Men and Avengers graphic novels.

Superheroes weren't the only ones. Early-twentieth-century posters of circus strongman acts presented curvy men with big bellies and solid limbs, with barely a chiselled abdomen, crisp bicep or hard pectoral in sight. As time passed, however, new representations of strength appeared in the form of the body builder, and with them the symbol of superhuman strength evolved from bulk to definition.

In an effort to make their muscles pop out, body builders – not to mention actors portraying superheroes on film – need to reduce as much tissue from beneath the skin as possible. That means they have close to zero per cent body fat, allowing every groove, every swell, every bundle of muscle fibres to present clearly through the layers of skin and spandex. Kritios Boy would feel right at home flexing his tiny but chiselled abs alongside Batman's own boy wonder.

Not that it matches reality. Superman might be an alien from Krypton, but those muscles aren't fooling anybody. One need only look to the Olympics to see how there is no single 'perfectly fit' human form. When tuned to long-distance running, swimming sprints, weightlifting or hammer throw, each athlete has the muscle size, bone length and fat percentage to suit their event's needs. Few, if any, look like Hugh Jackman as Wolverine; as it was with the Greeks, our archetypes don't always reflect reality.

Capturing perfect forms has never been about modelling nature, though. As has been the case throughout history, outward appearances communicate far more than just strength, grace or stamina. We use physical characteristics to portray attributes of health, fertility and economic value, and moral virtues such as persistence, moderation and personal pride. Muscles and symmetry are but proxies for status.

Looking 'good', whatever that might be, has always been about presenting as normal, reliable and an asset to the community – the very opposite of diseased. But just as our ideas about disease have changed, what 'good' looks like hasn't always been the same.

Until the early twentieth century, nobody in Western countries would have thought of darkening their skin through sustained exposure to sunlight. Suntans were associated with working long hours outdoors – something typical of those low on the social ladder, such as a labourer, fisher or farmer. To have light skin spoke of time spent indoors, and therefore of the wealth that permitted one to rarely step outside.

Similarly, in cultures across ancient China, white skin served as a sign of power and wealth. In modern India, skin-whitening creams are a booming industry to make the dark complexions associated with lower castes and racially vilified cultures look fairer. The centuries have seen all manner of toxic substances that mask and strangle the blood vessels beneath the skin come and go. The ancient Greeks and Romans, not to mention the pre-revolutionary French, used white lead to whiten their skin, adding just a touch of pink with red lead to accentuate a 'healthy' flush.

People have literally made themselves sick in order to look powerful, rich and healthy. When the famous eighteenth-century British beauty and socialite Maria Coventry passed away at just twenty-seven, it was more than likely the result of mercury slowly leaching into her lips and lead being absorbed by the pores of her face. And she was far from alone. In 1766 the English art historian Horace Walpole reported: '[T]hat pretty young woman, Lady Fortrose, Lady Harrington's eldest daughter, is at the point of death, killed like Lady Coventry and others by white lead, of which nothing could break her.'[42]

Nearly 100 years later, in 1869, the American Medical Association published a paper listing symptoms such as fatigue, nausea, headaches and paralysis, caused by the use of lead in a make-up called 'Laird's Bloom of Youth'.[43] Only in 1906 with the US *Pure Food and Drug Act* were laws introduced to control the use of potentially harmful materials in foods, cosmetics and pharmaceuticals. People knew there were risks, but subscribing to the illusion of health and power was more prized than the reality.

In most modern Western, Caucasian cultures, tans have become a sign of fitness, strength and vigour – of time spent outdoors in physical leisure. Now people risk melanoma by stretching out beneath UV lamps to achieve the 'healthy' glow of sun-touched skin. Fashion designer Coco Chanel is commonly blamed for the shift: the time she spent sunning herself in the French Riviera in 1923 encouraged her fans to swap pale skin for swarthier tones.

The shift in the implications of darker skin tones in Western populations coincided with changes in what it meant to be strong and powerful. Previously, leisure time was spent sitting and relaxing; today it's the time we're most active. Being tanned, slim and well sculpted is more than just a sign of good health and fitness; it's a visible display of the merits earned through enduring physical labour and imposing self-control, rewards that are portrayed positively in our art and choice of celebrity. Svelte and swarthy became the new pale and plump, as impressions of poverty shifted from thin, outdoor labourers to overweight, indoor office and factory workers.

For women through the twentieth century, a slighter figure has also become a sign of power and strength. Two world wars had revealed that without men around, women could more than just cope performing masculine roles – they could excel. The continued rise of feminism was accompanied by a rejection of the curvaceous, hourglass shape, in favour of less sexualised figures. For women, losing the bust and hips wasn't about being sexy; it was sex-empowering. From the flappers of the 1920s to the free-lovers of the swinging sixties, flat now spoke louder than fat.

If being trim is a sign of power and virtue in today's world, being big represents vice and weakness. The steady increase in body masses across much of the developed world has become such a concern that it's often now called an epidemic – a word that typically refers to the spread of a disease.

The health consequences of having too much body fat have been criticised since at least the fifth century BCE, when Hippocrates remarked on the Scythians: 'The men lack sexual desire because of the moistness of their constitution and the softness and coldness of their bellies ... In the case of the women, fatness and flabbiness are also to blame.'[44]

Plutarch noted, five hundred years later, that 'a thin diet is the healthiest for the body ... then ought the body to be light and in readiness to receive the winds and waves it is to meet with'.[45] Not long after, the influential Roman physician Galen wrote a description of how he helped a particularly rotund fellow lose weight by making him run until he sweated, and by recommending a diet that would fill him up and 'which afforded but little nourishment'.[46]

However, as Europeans approached the sixteenth century, any desire to be thin had faded away. So long as your weight didn't impede your ability to move around, a curved belly, full breasts and stocky thighs were all signs that you could weather a few seasons of famine. Curves didn't mean you ate cheap food or contributed little to society – far from it. Women were often described as beautiful when they had some junk in their trunk; on occasion, the ideal rump was compared to the flanks of a horse. The seventeenth-century Italian writer Fabio Glissenti recommended two different types of marzipan for Venetian and Neopolitan women, to help them keep a luscious size.[47] While there were pejoratives concerning gluttony, they tended to criticise taking more than one's share, rather than anything to do with appearance.

This isn't to say one couldn't be 'too big'. In the year of his death, William the Conqueror was so large he was accused by King Philip I of France of being pregnant. Not that the French were without their fair share of big monarchs – it was allegedly due to her repugnant weight that Philip's own wife, Bertha of Holland, was forced to give up her crown in 1092, even though the French king himself was also said to be too fat to ride a horse. These clearly weren't celebrated traits.

Illness from extreme obesity has never been a desired outcome – instead, it's clear that the line between what is valued and what is derided concerning body mass has shifted along with the cultural setting. Even in prehistoric times, carvings collectively referred to as 'Venus figurines' were chipped from stone, bone or antler, displaying the voluptuous forms of faceless women with wide waists and pendulous breasts. These art pieces can be dated to over 30,000 years ago, and have been found all across Europe, Russia and the Mediterranean. Who made them and what purpose they serve is anybody's guess, with hypotheses ranging from totems of fertility to representations of motherly deities, or even self-portraits by women looking down at their own bodies. But the fact they are so widespread, feminine and full-figured suggests an ancient respect for fat over thin when it comes to reproduction.

The modern beginnings of the obesity epidemic can be traced to the eighteenth century, when the English physician Thomas Short opened his monograph on the subject with the claim: 'I believe no age did ever afford more instances of corpulency than our own.'[48] Short advised picking a place to live that wasn't too damp and engaging in moderate exercise.

Less than half a century later the Italian anatomist Giovanni Battista Morgagni provided anatomical case studies of excess body fat, and of fat's relationship with the hardening of arteries, laying the empirical foundations for a relationship between different types of fat and health.

The year 1816 brought a milestone in the medical history of obesity, with royal surgeon William Wadd's publication of *Cursory Remarks on Corpulence or Obesity Considered as a Disease*, in which he warned that although progress has tended to 'banish plague and pestilence from our cities, [it has] also probably introduced a whole train of nervous disorders, and increased the frequency of corpulence'.[49] Wadd described the biology of fat, detailing its distribution, who it affected and its impact on disposition and health.

So great was the book's influence over the decades that followed, and the nationalistic fear in England that it had fatter citizens than those across the channel, that a dramatic cultural shift took place in how the English saw food. Diet and health had long gone hand-in-hand, to the point that it was all but impossible to distinguish between medicine and dinner for most of human history. The idea of modifying one's food intake as a means of losing weight had been around since the days of ancient Greece; however, the modern notion of 'dieting' – the widespread choice of foods based on a desire to avoid gaining weight – had its origins with the British undertaker William Banting. His 1864 booklet, *Letter on Corpulence, Addressed to the Public*, advised eating four meals a day consisting of green vegetables, meats, wine and fruit, without indulging in sweets, milk, butter and beer. It was so popular that dieting at the time was referred to as 'banting'.[50]

A smorgasbord of diets has since entered and faded from fashion, but perhaps none has been quite as unusual as that devised by the affluent American businessman Horace Fletcher, to whom is commonly attributed the mantra: 'Nature will castigate those who don't masticate.' His diet, which was less about the food he advised and more about the means of eating it, took off across the globe in the late nineteenth century. Fletcher believed that if it was to be digested properly, food had to be consumed in small bites and mixed thoroughly with saliva. In practice, this meant chewing ... and chewing ... and chewing food (and even liquid!) until it virtually trickled down the throat of its own accord. Like 'banting', the fad took on the name of its author, becoming known as 'fletcherism'.

Today, dieting is strongly linked with controlling the number of calories entering the body in the forms of lipids and carbohydrates, and matching it with the metabolism, which 'burns' the chemical energy as it reduces the compounds into molecules of carbon dioxide and water. While the term *calorie* has been used by chemists and physicists since the early nineteenth century as a measure of the energy it takes to heat one gram of water by a single degree centigrade, it only came to be associated with food in the 1890s, when an American chemist named Wilbur Olin Atwater started measuring the chemical energy released from food as it combusted. Based on his analysis of hundreds of food items, he suggested that if one wanted to lose weight, they needed to eat foods with fewer calories. His advice only gained popularity a couple of decades later, when a physician by the name of Lulu Hunt Peters wrote a bestselling book called *Diet and Health: With Key to the Calories*.[51] Dieting, fitness and weight loss is now a hundred-billion-dollar-a-year global industry, as waistlines continue to expand and fat is advertised as shameful, ugly and unhealthy.

One major issue with obesity being blamed on 'too much fuel' is that the bigger a person gets, the more energy they require. If pure gluttony wasn't the problem, as the Hungarian-born endocrinologist Hugo Rony presumed, something else had to produce abnormalities in appetite. His endocrinological hypothesis inspired a school of thought blaming obesity on hormones such as insulin, or overly enthusiastic lipid cells. For the first time obesity was being debated as a disease, and bad biochemistry rather than greed was being explored as a cause for at least some, if not the majority of cases.

A century of medical science has made it fairly clear that excess deposits of fat increase one's risk of developing various uncomfortable, debilitating and even life-threatening conditions, from diabetes to heart disease to cancer. But the numbers describing carbohydrates, calories, waistline centimetres, body mass kilograms and heart attack percentages don't necessarily tell a crystal-clear story. What exactly does it mean to be 'overweight' and 'obese', and what is the line between fitness and fatness? Putting it simply, when does being big become a disease?

To find an answer, we can go back to 1832, when a Belgian statistician named Adolphe Quetelet published an index of ratios comparing people's heights with their weights in order to study growth patterns. While he had no intention of using his data to sort the 'corpulent' from the healthy, his method was adopted a little over a century later, after World War II, and repackaged as the body mass index (BMI). The measure has become the standard for sorting those adults who are heavy because they have tall bodies from those who are heavy because they carry a lot of fat.

An individual's BMI is calculated by dividing their weight in kilograms by the square of their height in centimetres. So someone who is 200 centimetres tall and weighs 100 kilograms would have a BMI of 25 – right on the line between 'normal' and 'overweight'. A person who is just 150 centimetres but also weighs 100 kilograms would have a BMI of 44, well beyond the 'obese' BMI line of 30, and more than likely putting them into the category of 'morbidly obese'.

The statistics seem downright terrifying: across the Western world about two-thirds of adults are considered to be overweight or obese, according to their BMI. The exact number varies with gender and country, but not by much. As well, about a third of those who are overweight are actually in the obese category. If we zoom out to include every human on the planet, the WHO estimates that 1.9 billion people on Earth are overweight, and 650 million obese.[52]

While the index is a good way to make weight relative to height, and can offer a sweeping view of changing bodies across the globe, it suffers from a few problems when used as a simple shorthand for increased health risks. For one thing, BMI is hard to personalise because it doesn't take into account other features that contribute to your body mass, such as the amount of muscle you carry or the quantity of water you might retain. A short, young footballer might have a higher BMI but fewer health risks than a tall forty-year-old with a dad bod.

BMI also doesn't say anything about *where* the fat is carried, a factor that correlates with other health problems far more closely than simply how much fat you have. In 1947 the French physician Jean Vague noticed that, among his obese patients, those who carried their fat around the gut – in other words, who had large amounts of adipose tissue surrounding their abdominal organs – were more likely to develop cardiovascular disease than those with large amounts of subcutaneous fat.[53] Waist-to-hip ratios do a better job of predicting the risk of heart disease than BMI alone.

It's not that physicians don't take this kind of information into account when advising their patients. Many do. But the message people get outside of a detailed consultation inside the doctor's office is typically less nuanced. If we go by television shows encouraging us to lose weight, fad diets and gym memberships, celebrities lauded on magazine covers for their slimming down (and shamed for their plumping up), fat jokes in advertisements and sexy clothing on svelte runway models, the human body has an ideal shape and weight that we should consider not only normal, but also completely achievable – if only we have the willpower to eat fewer calories and exercise more.

Nowhere is this illusion of a universal cure for obesity made more glamorous than on the popular TV program *The Biggest Loser*. First airing in 2004 on NBC in the United States, the program has made a game show of weight loss, offering temptation challenges, presenting contestants with physical activities designed to tone and lose body fat, and eliminating those who don't lose enough weight by a specific time. The show has been a sensation in America, and has spread to over thirty countries around the globe, with programs produced as far abroad as Vietnam, Slovakia and Ukraine.

The show is promoted as inspiration for viewers to 'transform' their lives by sharing the stories of ordinary people with obesity as they lose weight, framing their journey from self-hatred to pride and joy over the course of each season as a struggle that can be managed by anybody with enough perseverance. The message is clear: being fat is a choice, and a moral failing. It's your fault. It isn't a disease.

In 2016 the US National Institutes of Health published their results of a longitudinal study on fourteen of the sixteen contestants in season eight of the US *The Biggest Loser*.[54] In the six years that followed the show's finale, every one of those studied – bar one – regained weight. Four became heavier than they'd been before they took part in the program. Anecdotally, reports abound of similar gains among former contestants around the world.

It's easy to assume that they all simply fell off the wagon for want of willpower. Who can sustain such a rigorous training regime, not to mention avoid the temptations of extra-cheesy-crust pizza and glazed doughnuts in the outside world? They were weak when they started, and they simply returned to their vices after the show, right? On closer inspection, something else seemed to be responsible.

The study showed that changes to the metabolism of the 'losers' could be more to blame. When the show started, each contestant's metabolic rate matched their size – big people have big bodies that burn a lot of calories. As the contestants burned away the kilos, the rate at which their bodies converted calories dropped. That wasn't unexpected, but what did come as a surprise was the fact that the drop in metabolism persisted.

Season eight's victor, Danny Cahill, went from a hefty 195 kilograms to 88 kilograms, losing over half his bodyweight. As the years rolled by, he put back on more than 45 kilograms, in spite of dieting and exercise. To avoid adding even more weight, he has had to eat 800 fewer calories per day than a man of equivalent size, possibly because his relatively brief time on *The Biggest Loser* slammed his metabolism. Far from being about just a matter of willpower and effort, Cahill, and the many other people whose weight exceeds what we view as normal, struggle against human biology, which evolved to prioritise building stores of fat.

As newsworthy as the single study's results are, it's important to note they aren't generally taken as a consensus, nor have they been solidly backed up by other research. The chief scientific officer of Weight Watchers in the United States, Gary Foster, has argued that the 'idea that the act of managing your weight and losing weight has somehow set you up to be in a worse spot' just isn't scientific.[55] That might be true to some degree, but there is a growing body of evidence that putting on weight in the first place – no matter what the initial driving force might be – alters our body to be more fat-loving.

A number of studies in recent years have found links between obesity and lower diversity of gut microflora, for example. Gastric bypass surgery that closes part of the stomach and reroutes the small intestine seems to shake up this microbial ecosystem by introducing new species.[56] The real kicker is that studies using mice suggest it's possible that such bariatric surgery mightn't even be responsible for the weight loss that follows – the new gut bugs could be achieving it all on their own.

What this means is that once we put on weight, losing it again might be made harder due to changes to our body's precious microbes. Like an infectious disease in reverse, we lose helpful microbes that would otherwise assist in managing the calories we absorb.

Weight gain not only changes our gut flora, it changes the very cells that store fat. Filling them to capacity can literally scar our adipose tissue, potentially making it harder to remove their contents.[57] To make matters worse, changes to the hormonal system that result from weight gain seem to add their own hurdles, ramping up hunger not just in the months after you start to lose weight but years later.[58]

It's easy for those of us who aren't overweight to imagine ourselves forgoing the extra slice of cake and walking off the burger, under the impression that weight loss is a simple equation of reducing calories. Really, it seems that fat deliberately makes it hard for a body to get thin, almost as if it has an agenda of its own.

Obesity isn't so much about broken biology as it is about biology behaving as it should in a world that is radically different to the one our ancestors lived in for millions of years. In some cases this leads people to conclude that the best diet is one that closely mimics the kinds of foods available to pre-civilised *Homo sapiens*. This is the claim behind the famous 'paleolithic diet', a food regime first proposed by a gastroenterologist named Walter Voegtlin in the mid-1970s, and given new life in 2002 when a nutritionist by the name of Loren Cordain wrote a bestselling book named *The Paleo Diet*. While Cordain holds a trademark on the name, online booksellers have literally thousands of book titles that involve some variation of diet based on what we think our ancestors ate, many of which are promoted by chefs and celebrities who lack backgrounds in anthropology or human evolution.

In essence, the diet advises consumers to avoid things like refined sugar, salt, cereals and legumes, and to eat plenty of nuts, grass-fed beef and seafood. While it isn't bad advice, and aligns fairly neatly with what most dieticians say about eating more greens and lean meats and fewer simple sugars and starches, the diet is based on the false premise that we can use a period in our evolutionary history to define what is normal – and therefore healthy – for our bodies.

Of course, we only need to look at the ratios of lactose intolerance across the world to see how the diets of pre-agricultural cultures reinforced our biology. But the fact that most Europeans have sufficient lactase enzymes in their digestive system as they develop past weaning age – in spite of such a recent reliance on milk from domesticated sources – is testament to the rapid degree to which humans have evolved to accommodate new diets. In fact, our diverse array of pre-agricultural ancestors picked and grazed from a variety of food sources across vast home ranges, a talent that demonstrated our resilience in the face of environmental change.

The many popular paleo diets out there often reflect very little about our true ancestral diets at all, thankfully. From what we can gather, ancient humans ate what they could, where they could, when they could. So their diets were varied, and had times of plenty and times of scarcity. Adding fat when you could was a good thing, and our metabolisms evolved to cope with the ups and downs of famine and feast. There were many days of without, interrupted by days of filling your belly followed by resting.

None of this is to say the consequences of weight gain shouldn't be a valid concern for individuals and society – not by a long shot. As our bodies store more fat around our organs, we can expect our risk of heart and renal disease, diabetes and cancer to rise, and as a society we need to consider what this means in terms of economics, personal freedom and individual welfare. Obesity is a growing social concern that needs a practical solution.

But a solution won't be found in the answer to the question of whether or not obesity is a disease. By asking that, we're reducing weight gain to a dichotomy between blame and biology, resulting in either the employment of shame, ostracism and fear in the name of a 'cure', or in the absolution of all responsibility. We reduce the issue of fat to its simplest components – based on an aesthetic of good looks – and thus avoid the harder question of how each individual can better understand their body to achieve good health.

Putting aside the morality of using misery to 'fix' a person's perceived failings, in practical terms the consequences of treating obesity as a simple case of weak willpower are counterproductive. The fear of shame has been shown to cause patients to delay health care that could provide assistance.[59] Physicians' judgements influence their treatment of patients, as they make assumptions about their health status that aren't based on evidence. One consequence is a dearth of data on how some pharmaceuticals, such as antibiotics or chemotherapy treatments, differ in their impact with size.[60] Being big just doesn't get you into most medical trials.

There's nothing wrong with chasing the perfect body. But when aesthetic ideals are mistaken for ideals of perfect health, and when they come to define the basis of disease, we need to seriously consider the reasons and methods behind our desire to help our loved ones.

# WEAK

In the 1980s, when the fitness industry began to boom to the tune of Olivia Newton-John's 'Let's Get Physical', and when legwarmers were briefly fashionable, my aunt and uncle owned a gym. What started as a small backyard enterprise grew into a warehouse-sized business, complete with a childcare area, weight training equipment and space for various classes. Oh, and it was painted bright pink.

My mother was an instructor at this gym, so during my teenage years I would stop by after school and wait around for a ride home. As it was a family business, I more or less had the run of the place, so would sometimes alleviate my boredom by flexing my adolescent muscles on the equipment or riding the stationary bikes, and would watch the variety of toned and flabby figures grunt and groan as they preened, perved and bulged in front of the mirrors.

I quickly learned that exercise is boring. Forget the 'runner's high' they talk about. For me, it just hurt, and it was ridiculously tedious. Worse still, it wasn't like a tooth extraction – one heave and it's done. No, this pain had to be repeated for any benefits to stick around. And if you stopped for even a few weeks, all that hard work slowly faded away.

Why bother? I was fine with sport, so long as there was some purpose ... as much as putting a round thing into some sort of net qualifies as a 'purpose'. Martial arts of various sorts were fun, and I liked swinging swords at people. Sure, like most teenage boys I would have loved a six-pack and chunky biceps, but after a few reps of bench press I convinced myself that cutting out one chocolate bar each week would be easier, and have much the same effect.

I knew then that pumping and jumping in gyms just wasn't my cup of tea. Still, if I wasn't sweating, I was behind the counter, where I'd file cards and serve protein bars. I'd see the same familiar faces stop by each afternoon to chase their ideal shape and drop their heartbeats per minute. I could tell from their files that a number of them had been in that morning as well. Five days a week, twice a day for one, two or even four hours at a time.

Officially, exercise addiction doesn't exist. According to the *DSM-V*, gambling can be considered a disorder if a person presents at least four criteria from a list of nine, such as being irritable if they cut down on gambling, or if they lie to conceal it. Kleptomania is in there, under impulse control disorders, similar to compulsive hair pulling and pyromania. But a compulsion to exercise isn't regarded as a mental health disorder in the same way.

On the face of it, why should it be? After all, there are no benefits to compulsively torching piles of old tyres, plucking out your arm hair or stealing hand lotion from the local 7-Eleven. But exercise is such a 'good' thing that it's hard to imagine we could suffer by overdoing it. While we're at it, eating fresh food is also a good thing, as is cutting your sugar and fat intake. Is it possible there's too much of a good thing?

For all the benefits you get by strengthening your muscles and regularly raising your pulse, exercise isn't without risk. The benefits of light exercise might outweigh the costs, but doing more doesn't necessarily mean more rewards to balance the risks. High-performance athletes, especially those who do endurance sports such as ultra-marathons and long-distance triathlons, suffer cardiovascular damage over long periods of time. A study of over 52,000 participants in an annual Swedish cross-country skiing event known as Vasaloppet found that male athletes with faster finishing times and more completed races also happened to have a higher risk of arrhythmia.[61]

The news for women who exercise excessively is particularly grim. Engage in anything beyond moderate exercise in the long term and you run the risk of your brain's hypothalamus misbehaving, screwing up your levels of gonadotropin-releasing hormone and preventing periods and disrupting your menstrual cycle. It can also trigger reproductive problems. Then there's the potential for hyperandrogenism (more 'male' hormones than 'female' ones), and hormonal changes that could mean you don't reach peak bone density.

That doesn't mean we should forget about exercise and sit safely on the couch; it means the benefits don't pile up faster than the risks as we push ourselves past daily jogging and weekly games of touch football. Quite the opposite. When it comes to exercise, there can be too much of a good thing, and some people can be addicted to the rush, much like those addicted to gambling or lighting fires.

There is no doubt that some of those friendly faces who habitually got high on the endorphins released by elevating their heart rates and adding a few more weights to their leg press were addicted, compelled by neurochemistry that craved the rush. By one estimate around 3 per cent of the general population could qualify as being addicted to frequent or excessive exercise, if we applied the same qualifications as other forms of impulse control disorder.[62]

But craving a so-called natural high is only part of the story, at least for some. A percentage of 'gym junkies' are no doubt also driven by a distinct impression that their body is imperfect: that more exercise, more muscle and a more constrained diet is the solution to changing a body shape they feel repulsed by, whether it's too fat, too bony or poorly proportioned.

Body dysmorphia disorder is characterised by a persistent, intense focus on anatomical features that sufferers perceive as being morbidly different. This could be a nose they see as crooked, a shape they think is unattractive, body hair they think is excessive – anything that leads to distress and gives rise to an obsession to change that feature. Exercise is one way people try to change their aesthetics; others regain control over their body shape by reducing their food intake, changing their diet or ensuring that what they eat doesn't stay in their digestive systems for long.

Eating disorders – controlling the nature and amount of food we eat to such an extent that it severely impacts on other aspects of our health – appear to have a loose relationship with exercise addiction. An estimated four in ten adolescents diagnosed with conditions such as anorexia nervosa or bulimia also present behaviours that could be interpreted as a compulsion to exercise, suggesting that the cultural ideals of body size can push otherwise healthy habits into pathological compulsions.[63]

In the past, malnutrition resulting from the avoidance of food was more about spiritual beliefs than anatomical ideals. The fourteenth-century Dominican theologian Catherine of Siena died at age thirty-three after a stroke that was probably connected to her abstinence from food, which she pursued in spite of instruction by her superiors to eat. An English physician named Robert Willan wrote about a 'remarkable case of abstinence' in 1790, in which a patient died after a three-month fast as a result of 'some mistaken notions in religion'.[64]

It was only in 1868 that the term *anorexia nervosa* appeared with a comprehensive medical description, thanks to another English physician, William Gull. While he remarked on its commonness among women, there was no mention of their desire to reduce their body size. Fasting and devotion to one's beliefs, whether in penance, humility or even blackmail, has played a central role in faith and political manipulation for centuries.

Extreme fasting and purging mightn't be all that new, but the relationship between eating disorders and body image has appeared in just the past half-century. The American psychoanalyst Hilde Bruch brought the modern relationship between abnormal fasting, purging and body ideals to the public's attention through her research in the 1960s and '70s, which was made famous through her seminal publication *Eating Disorders: Obesity, Anorexia Nervosa, and the Person Within*. Yet only in 1980 was body dysmorphia added as a factor to the entry on anorexia nervosa in the *DSM-III*.

Today, the cut-offs for the condition are a refusal to maintain body weight 'at or above 85 percent of normal weight for age and height', with an 'intense fear of gaining weight or becoming fat, despite being underweight', and are divided into categories that take into account binge-eating and purging, and food restriction. At its height, the condition was diagnosed in as many as 135.7 out of 100,000 adolescent girls, with the number rising steadily in the West over the last half of the twentieth century.[65]

Closely related to an abstinence from eating is the careful cultivation of a diet that we mistake for being clean, ethical and – most importantly – healthy. In today's age of raw, unprocessed, superfood and vegan diets, a desire to promote eating well can easily slip into harmful territory. A term has even been coined to describe it: *orthorexia*. Caught up in fashions that present a vision of the ideal body fuelled by an ideal diet – one free of intoxicating preservatives and processes that remove nutrients, and with clean produce devoid of infectious agents – it's not uncommon for some of us to unwittingly take in toxic levels of unregulated substances, miss out on key nutrients and swallow unsafe loads of infectious agents.

Fashion feels like an overly whimsical word to describe how we define health and disease. Our Instagram fad for advertising our meal of kale and quinoa and showing off our yoga poses might feel new, but it's merely a new format for an age-old human behaviour. Good health is defined as much by trends that sweep across the world as it is by scientific investigation. And in a world where even bad ideas are dressed up in words that sound scientific, telling fashion from facts can be hard.

In Eastern cultures, anorexia nervosa was an uncommon condition, and very rarely a result of body image. If anything, eating disorders seemed to be associated with perceived discomfort, such as a sense of bloating. Psychiatrist Sing Lee observed the rise and shift in anorexia nervosa in Hong Kong in the wake of a popular news story featuring the death of a young schoolgirl. In 1994, word spread of a teenager named Charlene Hsu Chi-Ying, who was 'thinner than a yellow flower', collapsing on a busy city street. Reporters dug into Western psychiatric texts to diagnose-by-media her condition. Anorexia nervosa, they declared, caused by an intense fear of gaining weight.

While Lee reported that he had once seen no more than one or two patients with an eating disorder each year, by the end of the decade that number had skyrocketed to several per month. The sudden rise also hit the press: 'Children as young as 10 starving themselves as eating ailments rise,' claimed the newspapers. By 2007 about 90 per cent of anorexia diagnoses were identified as being caused by a fear of being too fat. Lee told *The New York Times*: 'When there is a cultural atmosphere in which professionals, the media, schools, doctors, psychologists all recognise and endorse and talk about and publicise eating disorders, then people can be triggered to consciously or unconsciously pick eating-disorder pathology as a way to express that conflict.'[66]

From Lee's perspective, there was a monumental shift in what constituted a 'normal' eating disorder in Hong Kong at the end of the century. 'As Western categories for disease have gained dominance, micro-cultures that shape the illness experiences of individual patients are being discarded,' he explained. Not only did expectations about what constituted a normal body size shift with Western influences, the boundaries of the Eastern form of anorexia nervosa changed to accommodate this new abnormality.

The history of disease hasn't been helped by the fact that the scientific foundation of health and medicine has largely emerged from Western institutions. The results have been literally somewhat WEIRD — capitals intended — since nearly everything we know about how the brain works has been based on Westernised, educated, industrialised, rich, democratised research subjects. For most researchers, the easiest way to test a hypothesis has been to find willing volunteers outside your own building, and preferably they'll all be fairly similar in order to reduce unwanted variables. In the United States, about two-thirds of university-based psychology research is gathered from students in exchange for beer money or course credits. While this isn't kept secret, it's rare for the media sources to make a big deal out of it, meaning normality is determined by the functions of affluent, English-speaking undergrads.

It's all but impossible to define disease in a way that doesn't imply or explicitly rely on a standard set by what we think is common, average or traditional. There's a good reason for this – we need to agree on a set of standards if we're to operate as a society, and the most equitable way is to look to values that are prevalent among those nearest to us, or values that have stood the test of time within our community.

The problem isn't that we have a standard to help us prioritise treatment of our misery. It's that in determining what is normal, we're biased to believe that our point in time and our position among all humans is privileged. Our brains have evolved to cling to the familiar as right, and to see difference as something to shun – just in case. From the lepers of ancient times, to gays, to the obese, to the uncircumcised of today, 'unclean' is synonymous with the abnormality of disease, regardless of the risk of infectious agent.

In treating 'normal' as a biological law, we create a false dichotomy, arguing over whether things we find different are set in stone as diseases or whether they're infractions of choice, ignoring the fact that the standard of normal is itself a variable. Unfortunately, for some, the very effort of trying to achieve normality causes an intense level of discomfort, which medicine then strives to relieve.

In the words of Albert Camus, author of the existentialist novel *The Outsider*, 'Nobody realizes that some people expend tremendous energy merely to be normal.'

# V
# DEAD

# ROTTEN

I used to be a vampire.

Not the kind that sparkled, or avoided garlic. Drinking the blood I took from patients would have been a sackable offence, I'm certain. No, I was the kind of vampire who worked the graveyard shift in a hospital medical laboratory. By a quirk of cost-cutting I was also trained to collect samples from patients. There I was in my early twenties, a virtual stranger to daylight, sucking blood and analysing it on an assortment of clanking, buzzing, beeping machines. A joke was born.

Simply running diagnostic devices the size and shape of washing machines and counting cells down a microscope inevitably reduced the sick and the scared to little more than a barcoded tube. So I was privileged to meet the people who were in fear and pain, suffering a heart attack, pancreatitis, burst appendix, broken limbs, broken pelvises, dementia or overdose. The emotions on their faces revealed to me the lived experience behind the raw numbers, and so the evenings spent being a vampire all those years ago left an indelible mark on me.

A handful of patients' stories still stand out in my memory. There was Elsie, who in spite of her shattered hip wanted to tell me a dirty joke as a nurse struggled to take her ECG. I also remember an old alcoholic called Keith, who in moments of low lucidity sometimes thought I was his son.

Then there was Frank. To be honest, I can't actually remember if that was his name, but Frank is as good a name as any in a pinch, and I'm sure he won't mind. He's no longer around to argue. Frank, you see, didn't wake up when I gave his shoulder a gentle shake at 2 am during one particular shift.

Waking sleepers was as much fun for me as it was for them – not that all patients woke easily, especially those who were heavily medicated. But when taking their blood, protocol often demanded that I still first verify their identification and requisitions with on-duty medical staff. Some tests just can't wait for breakfast.

I set up my kit on the chair next to Frank's bed anyway, giving the poor guy another minute or so to sleep, and then snapped on my gloves. I nudged Frank again, introducing myself in less of a whisper as I wrapped a tourniquet loosely around his upper arm, in preparation for the withdrawal.

I can't tell you exactly why I suspected Frank wasn't alive. I remember his eyes were half-open in the fluorescent light, but that wasn't exactly unusual. His chest wasn't rising and falling, but again, such movement wasn't always obvious. Whatever it was that made me twig, I decided to check his wrist for a radial pulse, then pushed a fingertip against his carotid artery. I leaned in close to see if I could detect his breath on my cheek.

In those few seconds, I gradually became aware that I had just touched a corpse. The man who I'd thought of as a sleeping patient just moments earlier was now ... different. Frank's wrinkled skin wasn't blotchy or pale, his muscles weren't rigid, his body temperature was not yet cold, and he didn't smell. But something blinked in my mind and a knot twisted in my gut – Frank wasn't driving this cab anymore, and that made me feel an intense emotion I couldn't quite identify. It wasn't fear, nor anxiety. I didn't feel any dread, horror or even morbid curiosity. Deep down, I felt a mix of introspection and awe at what was my first close encounter with a deceased human.

In the words of Joseph Stalin, one death is a tragedy; a million is a statistic. The fact that a couple of human lives end on Earth almost every second or so is trivial until it's a life that ends in front of us. For me, death was an academic topic until I was touching Frank.

Ironically, our fight against the Grim Reaper is often indistinguishable from our battle against disease, despite the fact that there's nothing more functional for most organisms than eventual death. As a global phenomenon, biology has ticked on uninterrupted for several billion years, in endless cycles of reproduction. But as individual units, the fact that you and I will die is as predictable as tomorrow's sunrise.

Just as our concept of disease reflects the values of our particular culture, our personal relationship with death, dying and dead bodies also depends largely on our tribal affiliations. Just think about laying your hands on a corpse, for example. Do you flinch at the thought, or barely blink? Would you ever place your mouth on a cadaver's lukewarm lips? Snuggle up in bed with one? Or does the very thought of even being in the vicinity of a cold body fill you with dread?

For many of us, getting too cosy with a deceased human inspires an 'ick' factor. It's embedded in custom, laws and what we might think of as common sense. Dead, for many of us, feels dirty, and is aligned with stench, germs and rot.

Take the Judeo-Christian Book of Numbers, for example: 'The one who touches the corpse of any person shall be unclean for seven days.'[1] In some places around the world, those tasked with dealing with the dead are of a lower social ranking, if not outcasts. Handling dead bodies is often the job of specific castes, such as the untouchables of Hinduism and the *burakumin* of Japan, though in many places those who touch the dead are ostracised on account of their occupation alone.

If touching is bad enough, consider something more intimate – eating food that has been in contact with or near a body. *Sin-eating* is an old custom that has since evolved into a variety of forms, based on the belief that the living can take on the burden of a dead person's sins by consuming food or drink. The tradition can still be found in the form of a symbolic cake or beverage at a funeral or wake. Food and drink are either passed over a coffin or a body, or in some instances placed on the breast of the cadaver before being swallowed.

In the past, the poor sap who gobbled up a sinner's transgressions wasn't always a friend, or even necessarily well liked. A 1926 book on Victorian-era funeral practices explains:

> Abhorred by the superstitious villagers as a thing unclean, the sin-eater cut himself off from all social intercourse with his fellow creatures by reason of the life he had chosen; he lived as a rule in a remote place by himself, and those who chanced to meet him avoided him as they would a leper. This unfortunate was held to be the associate of evil spirits, and given to witchcraft, incantations and unholy practices; only when a death took place did they seek him out, and when his purpose was accomplished they burned the wooden bowl and platter from which he had eaten the food handed across, or placed on the corpse for his consumption.[2]

Maybe you're not uncomfortable with wolfing down a snack with a stiff, but when it comes to making a meal of the meat of an ex-human, it's a rare person who can stomach the thought. Cannibalism is a universal taboo, with very few exceptions.

Among the countless bones representing tens and hundreds of thousands of years of human evolution, only a handful of *Homo sapiens* and a few Neanderthals show any signs of being gnawed on by human teeth or expertly butchered for a fry-up. And it's hard of course to tell if they're evidence of diet or desperation. One solid example of cannibalism on the coast of Spain 10,000 years ago did align with global climate change at the end of the last ice age, but there also seemed to be a diversity of food resources in their surroundings at the time, making the 'fell on hard times' hypothesis less tenable.[3]

Most examples to the contrary usually prove the rule. Take the Donner-Reed Party, for example, who famously resorted to dining on the dead in the winter of 1846–47, having been trapped by the Sierra Nevada snows during a pioneer mission to California. Then there was the 1972 plane crash in the Argentine Andes made famous by the movie *Alive*, where surviving members of a Uruguayan rugby team resorted to cannibalism when staring down the barrel of starvation. In such cases, the stark choice between living and starving was necessary for them to overcome the anxiety of digesting another human.

As rare as the practice seems to have been in the past, and it's virtually non-existent today, there are occasional stand-out examples of culturally sanctioned cannibalism on record. Anthropologists break consumption of human flesh into two types: *endocannibalism*, meaning eating those you know, which is typically done to honour the dead, and *exocannibalism*, which involves eating those you're not overly fond of, such as an enemy or member of another community.

Gruesome rumours of 'man-eating savages' don't always provide the most reliable anthropological descriptions, but these stories have at least hinted at cultural practices that scholars deem plausible. Stories and woodcuts of South America's Tupinambá people in the sixteenth century range from the brutal to the fantastical, presenting them as a dog-headed people who eat the meat off the bones of any Spaniard they capture. But even the reasonable attempts to humanise Tupinambá culture have done little to refute a reputation of occasionally butchering and eating an enemy. The long vanished culture left little behind to inform insights into their customs and habits, other than the records of a sixteenth-century German traveller named Hans Staden. According to Staden's own written testimony, he was held captive and feared being eaten on account of being an enemy of the Tupinambá.

A more recent example can be found in the indigenous Brazilian people called the Wari', whose past cannibalistic practices sat at the other end of the spectrum. When kin passed away, Wari' people would mourn for several days. Just as severe decomposition set in, they would cook and shred flesh taken from the body and place it on woven mats. Distant family members would be invited to consume the meat, using long skewers to avoid the disrespect of touching the flesh with their hands.

This wasn't a pleasant meal, like eating freshly hunted game; the food was on the verge of rotting, after all. Children might also be invited to join in, usually to consume the deceased's brain. Any remains that weren't consumed were often buried, though there are also records of the bones being crushed up and mixed with honey, and then consumed.

The funerary practices of the Fore people of Papua New Guinea could give us a glimpse into why the custom of roasting up grandma after she's breathed her last has never taken off generally. Arguably in the belief that it freed the deceased's spirt, close kindred would consume the body, with women and children traditionally consuming the brains, much as the Wari' did. In the mid-twentieth century, anthropologists recorded a condition among the Fore they called *kuria* – or *kuru*, as it became known to Western medical researchers. The disease is now understood to spread via consumption of body tissue infected with a rogue protein, and it causes shaking, an inability to maintain posture and eventually death. The protein, called a *prion*, bends similar proteins in the host's brain, behaving a little like a virus in how it is transmitted.

Though it's passed on through infected material, kuru isn't a disease transmitted solely by dead bodies. Nonetheless, many of us still retain some vague sense of a body turning infectious purely on account of no longer being alive. The sense that the dead are agents of infection is so powerful that it often impacts on the management of large-scale disasters. Following Turkey's devastating earthquake in August 1999, the nation's government requested tens of thousands of body bags be imported, fearing the 17,000 bodies pulled from the rubble and being held in makeshift morgues might give rise to an epidemic of disease.[4]

In 1869 physicians at the second annual International Congress of Medical Sciences heard a talk on the benefits of incinerating the dead over burying them, 'in the interest of hygiene'. This sparked a movement to not only change laws on how to deal with the dead, but also a wave of experiments aimed to improve cremation methods.

Opting to be reduced to ashes after death is now a popular choice for many of us, but is it really better for public sanitation? Population movements in the wake of devastation and the contamination of drinking water and food supplies can easily spread pathogens, but decomposing bodies – while they may be uncomfortable to see or smell – don't pose much of a risk. For one thing, most of the body's microbes don't survive long without our protection, so microflora that are released into the environment aren't generally a problem. Fluids contaminated with pathogens that are easily transferred through bodily fluids – such as those causing Ebola, smallpox or hepatitis C – certainly count as a biohazard. But being dead doesn't spontaneously turn a person into a vector of disease.

Knowing this makes little difference to most of us, though, as being repulsed by decay of any kind is a deeply entrenched behaviour. After the Turkish earthquake, the WHO advised the government that workers should prioritise devoting resources to providing clean water and caring for the injured where possible, as uncomfortable as it might be to leave the dead exposed to the elements.

While it's common for us to prefer to keep our distance from the recently departed, there are diverse and colourful cultural exceptions. In the remote highlands of the Indonesian island of Sulawesi, many of the indigenous Toraja people still follow a tradition of keeping the deceased in the house for weeks, months or even years, with funerals delayed so relatives have time to journey from far afield to visit the body.

Historically, the people responsible for turning cliffs into the ancient city of Petra in what is now Jordan also lived as relative neighbours with their dead. Over the course of several centuries, the Nabataeans carved a city out of the beautiful rose-gold sandstone in the desert wilderness. Much of the ancient city has been reclaimed by the desert sands; what stands out are hundreds of doorways carved into the cliffs, each the entrance to a small room that once housed the dead. Without much in the way of bones to analyse, archaeologists can't say what kinds of funerals the Nabataeans gave, but by the early second century CE – just decades after the city fell under Roman rule – the dead were no longer housed in nearby apartments overlooking the city. The new citizens weren't as comfortable living cheek-to-cheek with interred bodies. Petra and its culture would persist through earthquakes and even the fall of the Roman Empire, but the dead would never again be kept inside the city limits.

Putting our kin into a hole after they've passed away clearly isn't a recent ritual in human history, even if nobody knows precisely when or why our ancestors first took an interest in hiding their dead. More than likely, it has its roots in a deep-seated social behaviour; elephants pay close attention to their own species' bones, after all, and regularly mourn the loss of herd members.[5] Primatologists have recorded interactions between living members of ape communities and their recently deceased kin, such as grooming of the corpse and squabbling for access to the body.[6] But there is no evidence of a non-human ever interring or destroying dead members of their population in a ritualised, mournful fashion.

Our relationship with death might be unique among animals, but it still potentially dates back hundreds of thousands of years, and could be a behaviour we share with at least some of our closest extinct cousins. In 1997 a team of palaeontologists brought up thousands of bones from a deep pit inside a cave in northern Spain's Atapuerca Mountains, a discovery that gave the site its name – Sima de los Huesos, or the 'Pit of Bones'. The age of the remains was determined to be somewhere between 350,000 and 500,000 years, putting them in the latter half of the Pleistocene, or what we colloquially call the Ice Age.

A number of the pit's bones were of cave bears, but the bulk were left by twenty-eight human relatives of various ages. To be precise, these relatives belonged to the species *Homo heidelbergensis*. All but 10 per cent of the world's *H. heidelbergensis* fossils come from this single haul, in fact, providing us with a grim snapshot of one of our species' most recent ancestors.

It's a discovery that carries an intriguing question. What were so many people doing down a hole? Half of them died between the ages of ten and eighteen. Only one was over the age of forty. One child's skull showed signs of a growth deformity, but most seemed relatively healthy. It's unlikely the group was living down there, since there were no other signs of habitation. The only artefact was a single hand-axe, a tool that would undoubtedly be of great value to an ancient community, so wouldn't have been lobbed down a hole without good reason.

One possible explanation is that the remains had been washed into the pit from elsewhere; an absence of smaller bones at the site hints at the possibility of a chaotic re-assortment. But another hypothesis stirs the imagination far more: the bodies had been deliberately entombed, perhaps as a means of burial. If that's correct, it's the oldest tangible example we have of a species intentionally burying their dead.

*H. heidelbergensis* were most likely a fairly clever folk. Their average brain size was slightly larger than ours, and they've left examples of fairly advanced tool production. But the question of what they did with their dead is a hard one to answer with confidence. If it turned out that the pit of bones wasn't a tomb, our cousins the Neanderthals could provide the only other example of what is described in anthropology as *patterned mortuary behaviour*. Even this is kind of controversial. What qualities do fossilised remains need in order to describe some form of death ritual? Hints could be found in the proximity of artefacts, or in the position or location of the body, but there's still plenty of room for argument.

Sima de las Palomas, in south-eastern Spain, is the final resting place of three adult Neanderthals, whose 40,000-year-old skeletal remains were found beneath heavy boulders that appear to have been dropped from some height.[7] Palaeoanthropologists debate whether the rocks were a cause of death or deposited post-mortem. The fact that the bodies' arms are folded near their heads might hint at ritualistic behaviour, as might the discovery of charred cat paws nearby.

A site in Iraq offers another tantalising suggestion of Neanderthal mortuary behaviour. Of the nine Neanderthal bodies discovered inside Shanidar Cave, all of which were dated to about 60,000 years old, one was found to be dusted with pollen, suggesting the possibility of plant material (such as flowers) being laid over it. Unfortunately, researchers can't rule out contamination, so again there's plenty of room to debate whether these bodies were ritualistically laid to rest or simply removed from sight.[8]

To current knowledge, the oldest confirmed burial took place in Australia some 42,000 years ago at what is now known as Lake Mungo, in New South Wales. There's no longer any water in the dry, salty depression, but before the end of the last ice age it would have been a virtual paradise, one where massive wombats and huge lizards trundled across the landscape. The ancestors of today's Indigenous people not only made use of the area's bountiful resources – there's evidence that the remains of a male skeleton were dusted with red ochre – they also ritually buried their dead.[9]

Perhaps coincidentally, another set of remains in the same area, dated to roughly 25,000 years ago, detail one of the world's oldest examples of an intentional cremation. The remains of a woman appear to have been incinerated, broken up, burned a second time and then covered with ochre.

Before the 42,000-year-old bones of 'Mungo Man' were found in 1974, human occupation of Australia was reckoned to be only around 20,000 years. His discovery not only caused a rethink of how ancient humans dealt with the dead in complex ways, but challenged ideas of how long Indigenous Australians had been on the continent. Today the Indigenous cultures of Australia can trace their occupation back beyond 65,000 years.

Mungo Woman's ashes were repatriated in 1992 to the traditional owners of Lake Mungo, the Ngyiampaa, Mutthi Mutthi and Paakantji people. She now rests in a locked vault. Controversially, until 2017 her counterpart's bones lay in a box roughly 500 kilometres away, at the Australian National University in Canberra. Now Mungo Man rests close to where he was found, but ancient remains across the world are still tucked away in museum draws and laboratory cupboards. A tension exists between the academic value of such ancient relics and the emotional value imbued by those with a connection to people they regard as family. One person's fossil is another's grave; in some cultures, even the passing of tens of thousands of years doesn't erode people's sense of connection with their dead ancestors.

For practical purposes, disease isn't a purely biological phenomenon, and neither is death. Our attitudes towards the process of dying, its sense of inevitability and its risk aren't universal. Nor are the ways we deal with the physical remains of our loved ones after they've breathed their last. It's easy to see why we, as social beings, would continue to feel an emotional bond with the body of a friend or relative, but what exactly prompted our ancestors to dispose of their dead in the first place, all those thousands of years ago?

# EXPIRED

From the comfort of civilised society, we might think it's clear why we have funerals: who wants rotting corpses and scattered bones lying around their streets? No doubt that's an incentive to not give up such a longstanding tradition, but it doesn't explain why our ancestors went to the effort of burning or burying a body, especially among communities that could simply walk away, as most other animals do.

Today's rural migratory communities could provide a glimpse of what many wandering cultures might have done in our deep past, especially when burial or cremation was inconvenient. Mongol clans, long before the cultural change brought about by the People's Revolution in the early twentieth century, practised what was referred to by some as *ködägäläkü*, or 'placing a body on the steppes'. Similarly, in Sichuan, China, dead bodies are left out for carrion animals to break down, after which the bones are retrieved and reduced to splinters with a mallet, then fed to other animals.

Leaving the dead in the elements to be broken apart was no doubt easier than burying them in stony, often frozen ground, and with limited supplies of wood, getting a hot enough fire going for a cremation was difficult. There are examples of indigenous American customs, too, in which bodies are wrapped and placed on scaffolds in the open air for a time, possibly as a way of keeping them away from ground predators such as wolves, while still leaving them open to nature.

But even if a corpse isn't reduced to ashes or put underground, there isn't a single human culture on Earth today that doesn't deal with the remains of its dead in some kind of ritualised fashion.

Nobody leaves their loved ones out for the ants, at least if they can help it.

If dead bodies don't typically infect us with disease, sanitation is an unlikely explanation for the first funerals. It's possible they might have been conducted to keep the dead from attracting scavenging predators, but this doesn't even begin to explain the origins of other funerary customs. Palaeolithic graves commonly contained more than just bones, after all, with the most ancient often marked by dustings of ochre or plant materials for decoration; as time progressed, we see food items, personal possessions and even expensive ceremonial objects.

Such behaviour raises a lot of questions, and we have so little to go on to get any answers. Only with the advent of writing did we gain some insight into what any of our ancestors were thinking when they buried their dead: by the time we first scratched symbols into clay, it seems we were already trying to understand the deeper meaning of life and death. Tablets displaying one of the earliest forms of script, created 5000 years ago by the Sumerian people in Mesopotamia, record transactions, poems and stories. The famous Epic of Gilgamesh tells a story of the gods and their efforts to make the king of the city of Uruk humble as he sought the gift of immortality. Importantly, it provides us with one of the oldest examples of an explicit belief in a transition between life and an existence that follows, albeit one painted as a dismal place somewhere underground:

> Seizing me, he led me down to the House of Darkness, the dwelling of Irkalla, to the house where those who enter do not come out, along the road of no return, to the house where those who dwell, do without light, where dirt is their drink, their food is of clay, where, like a bird, they wear garments of feathers, and light cannot be seen, they dwell in the dark, and upon the door and bolt, there lies dust.[10]

To the Sumerians, life without a body wasn't at all pleasant – at least, not according to this story. The dead are pitiful versions of the living, without direction, hope or ambition; they eat and drink dirt and dust. More importantly, they are entities that invoke fear, and would eat the living if they were to ever return.

Among the Sumerians and other Mesopotamian cultures, such as the Akkadians and Babylonians, we find the first recorded efforts to comprehend a physical difference between life and death. The Akkadian word *napistu* means both 'breath' and 'life', the oldest implicit recognition of life's dependence on air. This relationship is also expressed in the Chinese word for air and spirit – *qi* – and eventually our own word *spirit*, with its etymology in the Latin word *spirare*, also meaning 'to breathe'. Not only is the presence of breathing a sign of living, but the insubstantial quality of the air we rely on could poetically reflect the inchoate, intangible nature of life. Either way, it's not hard to see why breath is a recurrent metaphor for life across the ancient world.

From a handful of Neolithic clues we can work out that even as we recorded our philosophies in detail for the first time in history, we were already distinguishing between the meat of our bodies and an intangible essence we think of as the human mind.

The Egyptians put a lot of thought into the fundamental nature of human existence. Used as far back as the sixteenth century BCE, a funeral scripture called the *Book of the Dead* (or, more accurately, as it is called in many older examples, the *Book of Emerging into the Light*) detailed precisely what to do to help the departed keep hold of their personality. With details that evolved over the generations, the book instructed priests on how to ensure a person's core components stayed intact after they died.

The mortuary ritual most of us associate with ancient Egyptian cultures is, of course, mummification. The term *mummy* wasn't coined until Egypt came under Arabic rule in the seventh century CE, from a Persian word for 'bitumen', thanks to the blackened state of the preserved bodies. The Egyptians called these preserved remains *sah*, as distinct from the living body, which they called *khet*. Mummification was practised in Egypt as far back as 4000 years ago, and spread throughout the Mediterranean to be adopted by many Greek and Roman communities, beginning to fade in popularity only around the first century CE.

Why put so much effort into staving off corruption of the physical body? It wasn't because the ancient Egyptians believed they could use it to walk around again. Yet the physical body was still considered an essential part of existence. It was the recognisable part of an individual, like an afterlife ID card.

The heart was believed to hold memories and thoughts, so it was kept in place in the chest. A spell was inscribed on a carved iconic scarab to make sure the heart didn't betray its owner when it came time to be judged. Other organs, including the digestive system and the liver, were removed and stored nearby in jars. Spells were then read from traditional texts to ensure the body kept its basic integrity. Strange as it seems now, the brain wasn't worth much, so it was whisked into mush and poured out through a hole chiselled in the nasal cavity.

In addition to the *sah*, the body's life force – or *ka* – also required safe-keeping, and that was accomplished by reciting the right spells. Another kind of soul-like essence called the *ba*, on the other hand, was free to wander far and wide after death, so it could meet and greet the gods. It was often depicted as a bird with a human's head, symbolising its freedom from the limits imposed by the body. Then there was the individual's name, and lastly their shadow, or *shut*, which was also glued into place with incantations.

With all their personal components secured – their body, their life force, their soul, their name and their shadow – a dead person could live on with the gods in the form of an entity called an *akh*, so long as their heart didn't decide it was a good moment to get all anxious about that night they two-timed their wife after a few too many beers one Midsummer Festival on the final moment of judgement.

To the ancient Egyptians, the heart's steady rhythm was directly responsible for driving the limbs – the heart was not so much a pump as a vessel for consciousness. It stood to reason, then, that on death the heart could be interrogated by the gods, and would indicate one's moral worth. In the hall of Ma'at – goddess of truth and justice – the organ was placed on a set of scales against a feather. Should the heart sink, it would be judged as heavy with guilt. Potential sins were listed in the papyrus of Ani – they included theft, prying into others' matters, making another cry or committing slander.

So the Egyptians had a long, detailed book of rules describing one's passage into the afterlife. The Sumerians, by contrast, had: Step 1 – die. Step 2 – wish you hadn't; it sucks being a shade.

Why did Egyptian culture develop such an intricate system to explain death? One hypothesis is that their increasingly complex social system – with its development of a dynastic state, a developed middle class, and an ever more detailed bureaucracy – bled into their perception of an afterlife. Funerary practices changed significantly with the fall of the Naqada culture, which preceded the first rulers of a somewhat unified Egypt. As the society became more hierarchically structured, so too did their beliefs concerning individuality and humanity.

A similar change took place a few thousand years later around the Aegean, which happened to be a mosaic of dialects and cultures. There was no single state, no god-king, no detailed bureaucracy, but instead a relatively tribal system of rule by a governing class. This was the world of Homer, and it was more like that of the Sumerians than Egypt's more formal society. It's not all that surprising, then, to find similar accounts of the afterlife in the poet's works.

In the *Odyssey*, Homer writes of the dead hero Achilles expressing to Odysseus how he'd rather be a poor living peasant than a dead king in the afterlife.

> I would rather serve on the land of another man
> Who had no portion and not a great livelihood
> Than to rule over all the shades of those who are dead.[11]

Unlike the complex bits of the dead described by the Egyptians, Homer's idea of personhood was of something more like Gilgamesh's pitiful shadows. Referred to as a person's *psyche*, this ghost-like presence is both a life force and an immortal mind that journeys into another realm. The word's roots lie in the term *psychein*, which once again is associated with the act of breathing. Homer wrote in the *Iliad* that the life force 'cannot come back again ... once it has crossed the teeth's barrier'.

As with ancient Egypt's political changes, the Greek philosophical revolution gave rise to a new set of thinking tools to describe life and death. Inspired by new systems of economics, academics interpreted nature by way of universal, impersonal laws. The fifth-century-BCE philosopher Democritus, who was among the first to see matter as comprised of tiny units, which he called *atoms*, proposed that the motion of our bodies was sparked by a force comprised of the essence of fire. For Democritus, our thoughts and memories ended with our bodily functions, and our mind decayed like the rest of our flesh. He saw no room for *psyche* or spirits.

A few decades later, Plato envisioned a realm of nature that was invisible to our corruptible senses: a system that was purer and more perfect than what we could perceive directly. In this space, objects existed as forms that gave rise to their equivalents in our tangible realm. For example, an archetypical 'Platonic' sphere was perfectly round, unlike oranges and lizard eggs in the real world, which are only very nearly round. A person's soul was also a Platonic object, capable of interacting directly with other pure forms through thought, and imprinting them directly in our mind's eye, unfettered by our meaty senses of skin and eyeballs.

This was the role of reason – to connect a person's soul with the universe's forms (things like numbers, perfect circles and mathematical relationships), and see them as they truly were. Direct observation was fallible and susceptible to all kinds of illusion; by thinking rationally and logically, our minds could 'see' what our eyes could not.

Plato and his contemporaries transformed ancient philosophy. While those before him also understood the limits of perception, his school turned the conflict between what we directly sense and what we can logically reason into a feature of nature. This so-called metaphysics was like physical reality, in that it existed independent of our imagination, but like the things in our imagination, it was more ideal than tangible.

Plato's student Aristotle built on his teacher's idea of a metaphysical landscape to explain natural phenomena. His prolific writings passed through the hands of future philosophers, scientists and theologians in the Near East and Middle East, and through Spain into Christendom, inspiring centuries of scientific and religious musings on how life differed from death, and how the mind differed from matter.

For millions of modern Christians and Muslims, souls, heaven and eternal life would not be the same if not for Aristotle's deliberations on life and death. Centuries of theological thought have combined with ancient Greek philosophy to give us the immortal, immaterial mind. Yet centuries of scientific thought have also combined with Greek philosophy to give us the mortal, mechanical mind as well. The two don't reconcile very well – not that some great thinkers haven't tried.

Like most educated folk of the early twentieth century, the American physician Duncan MacDougall was a keen believer in the existence of the soul. As a student of medicine, however, he concluded that whatever gave us life had to somehow subscribe to the same laws of physics as everything else in nature:

> If personal continuity after the event of bodily death is a fact ... then such personality can only exist as a space occupying body, unless the relations between space objective and space notions in our consciousness ... are entirely wiped out at death and a new set of relations between space and consciousness suddenly established in the continuing personality. This would be an unimaginable breach in the continuity of nature.[12]

At 5.30 pm on 10 April 1901, MacDougall commenced a grand experiment. A man sick with tuberculosis was placed into a bed on a set of finely tuned scales. The illness had progressed to the point that death was evidently near, and at 9.10 that evening he passed away.

The good doctor noticed that during those hours he had to regularly adjust the scale by a fraction of an ounce so that in all, the dying man appeared to lose just over 28 grams per hour. MacDougall blamed this on moisture lost from the dying man's sweat and breath.

> At 9.08 pm, my patient being near death, for the last time I set back the shifting weight on the beam so that for the last ten minutes the beam end was in continuous contact with the upper limiting bar. Suddenly at 9.10 pm the patient expired, and exactly simultaneous with the last movement of the respiratory muscles and coincident with the last movement of the facial muscles the beam end dropped to the lower limiting bar and remained there without rebound as though a weight had been lifted off the bed.

MacDougall believed he had captured the moment a physical soul left the body. It took the 'combined weight of two silver dollars' to reset the scale. That's three-quarters of an ounce, or 21.3 grams.

Ever the scientist, however, he knew his conclusion warranted testing and retesting. He considered the variables, took into account the error margin of his scales and even experimented on dogs (which produced no equivalent change in weight). All the while, he endured the criticisms and opposition of those offended by his efforts. MacDougall was so wary of the impact of his work that he refused to publish on it. Only when rumours slipped into the media did he feel obliged to present his findings to the public.

MacDougall wasn't the first, or the last, to attempt to describe the physical connection between the living and the dead. Since his experiments, a handful of similar studies have endeavoured to find a material theory of the soul; most have focused on animals. In the 1930s, for instance, an American professor by the name of Harry LaVerne Twining compared mice killed in open beakers and mice killed in sealed containers, and found a change in mass coinciding with the presumed moment of death only in the open conditions.

For centuries, if not millennia, cultures across the globe have debated the relationship between biology and chemistry, and between consciousness and physics, using a mix of natural philosophy and religion. Death feels significant, and humans have come up with innumerable ways to try to explain why.

Once, anthropologists suspected that the evolution of formalised religion – as opposed to an animistic sense of human-like awareness in our environment – was a consequence of settling down during the agricultural revolution. It makes sense, with organised systems of ritual and priest castes seeming to emerge from the division of labour and surplus resources, not to mention greater investment in natural phenomena to not curse a land with famine. But the Palaeolithic site of Göbekli Tepe, near Turkey's border with Syria, challenges that idea.

What was initially thought to be a medieval graveyard has since turned out to be the oldest example of a temple on Earth. The buried stone monuments, arranged in circles and carved with symbols, pre-date Stonehenge by some 6000 years, providing an early example of permanent constructions before humans were planting crops and living in cities. While Indigenous Australians were arranging rocks into fish traps and kangaroo corrals long before this time, the purpose-built structures at Göbekli Tepe seem to have been made only for mystic reasons.

Human remains have been found at the site among a jumble of animal bones, carved idols and mounds of earth and rock, but the disarray has made it impossible to say much about how they got there. In 2017 anthropologists published findings on several skull fragments that had been intentionally scored and drilled, as if to create grooves for string to hang or hold them in place.[13] This evidence of a 'skull cult' puts the temple into a new, darker light. It also tightens the links between religion and death: for as long as we've been building monuments to spiritual forces, we've honoured the journey of the dead in a designated spot.

Dying horrifies and intrigues us all at once. Those of us who don't believe in an immortal mind or metaphysical forces still find ourselves in awe of the non-living human body, as if the very reminder of our human mortality might somehow hasten our own end. I'm an atheist, yet custom and respect for the dignity of my loved ones means I would still cremate or bury their remains. Souls have nothing to do with my approach to mortality.

Cultures across the ages have crafted reminders of death – art pieces called *memento mori* – to emphasise our finite lifespan, and to encourage us to appreciate every moment we have. While we don't like to think about dying, we do like to remember how precious – and tenuous – life can be. So strong is this sense of finality that we fight to preserve it at all costs, sometimes regardless of whatever quality of life we might be forced into, or of the fact that different individuals and cultures value death and dying in different ways.

In many ways, death itself is a pathology. Like disease, its biological reality is framed within an ever-evolving culture, one that is also viewed through a lens of morality.

# ARRESTED

Right now, my heart's muscle tissue is coordinating a rhythmic wave of electrochemical tides that cause the muscle cells to contract and relax every half-second or so. This steady pulse is maintained by nodes of specialised muscle cells called pacemakers, which in turn are under the careful watch of my central nervous system to ensure they don't slip out of line.

At some point in my life – far in the future, I hope, though one can never tell – the conductor will drop its baton and the orchestra of cells will lose their collective beat. My cardiac tissue will quiver pathetically as the fibres of my heart muscle contract and relax without consensus, no longer working as a team to shove litres of blood through a multitude of narrow channels. This arresting of my heart's steady pulse will be labelled clinical death.

For most of human history, the absence of a clear heartbeat marked a finality to life. Without intervention, a heart that has stopped pumping has between a one in ten[14] and a one in twenty[15] chance of returning to its usual rhythm. Being lucky enough to have a helping hand pump up and down on my chest, performing cardiopulmonary resuscitation, might buy some more time for my heart to beat properly again, depending on how well the helper knows what they're doing. But even then my chance of making it to another breakfast table only rises to between 20 and 40 per cent.[16]

Without a heartbeat, my body's transport system swiftly grinds to a halt, causing a physiological famine of essential materials such as oxygen, glucose and a soup of hormones necessary for coordinating an array of tasks. Chemical by-products accumulate as they aren't whisked away, corrupting my body's pH and choking vital metabolic processes. Lowering my temperature might help slow the rate at which my nutrients are used up and wastes spill out, stretching out the time it would normally take for my biochemical machinery to grind to a halt, but short of achieving perfect stasis, eventually every cell in my body will succumb to the effects of starvation, choking and suffocation.

Some of my tissues will have a harder time coping in the chaos than others. Within ten minutes of losing oxygen, enough neurons in my brain's hippocampus will be so badly damaged that a return to normal neurological function is unlikely.[17] For comparison, amputated limbs can often (depending on their temperature) be reattached six to twelve hours after the blood ceased flowing, with skin and muscle tissues usually recovering and regaining satisfactory functioning.[Δ]

Even in those first minutes and hours after the oxygen runs out and carbon dioxide rises, my body will behave in ways that to any onlooker might seem to be a flicker of life begging to return. Random discharges of nervous activity have been known to produce a startling post-mortem phenomenon known as the Lazarus reflex, causing the deceased to raise their arms and drop them across their chest. I look forward to performing that party trick.

Ripples of neurological activity have been seen surging across the brains of beheaded rats a full minute after decapitation, hinting at some kind of desperate, coordinated panic as life steadily winds down.[18] Observations of brainwaves similar to the ones we have when we're asleep were spotted in a patient ten minutes after he was officially declared deceased by physicians in a Canadian intensive care facility in 2017.[19] Far from a drop-off, the death of even our most delicate of organs is a series of steps down a staircase of oblivion.

Without a continual supply of blood, eventually each and every one of my body's cells will lose their ability to fulfil their

---

Δ  There's lots of variation in the length of time you can have a part of your body detached. Temperature makes a big difference – so keep that limb chilled! (See Maricevich, M., Carlsen, B., Mardini, S. & Moran, S. (2011) Upper Extremity and Digital Replantation. *Hand*, vol. 6 (4), pp. 356–363.)

responsibilities. Some will sacrifice themselves, opening tiny pockets of suicide enzymes that dismantle their machinery until only simple organic remnants remain. Most will retain some integrity only until the ravages of oxidation, desiccation or decomposing microbes dissolve their structures and scatter their organic threads to the winds.

RIP, Mike McRae.

Still, life in my body stubbornly ticks on for a little longer. Without my biology to keep them in check, my magnificent ecosystem of microorganisms will tear down barriers, opening the larder for other microbes to feast upon. My guts will swell with gases and warmth as bacteria leak into my body cavity, spreading and breeding. Weeks after my final heartbeat, my core will still be tepid with the life that once sustained me, a bacterial uprising like my very own zombie apocalypse.

Finding a precise moment that confirms me as dead won't be as clear-cut as we might expect. Far from a sudden stop, life's inertia can takes minutes, hours or even days to come to a full rest, depending on where we draw the line between life and death.

When we talk about human life, it's rare that we limit ourselves to a general assessment of metabolic activity. As a person, I'm summed up by my brain's ability to construct and communicate awareness. Only when my brain can no longer channel electrochemical signals across its network of neurons will I come to my official death. From there, any ability to narrate my own story will end, and any anticipation of some ongoing saga fades. The permanent loss of my mind will be a far more significant interpretation of death than my quivering cardiac tissue or failing organs.

Introduced as a concept in 1965, the medical (and often legal) end of a person's life occurs when the brain is silent enough for it to be presumed to be incapable of producing a conscious signal ever again. In most jurisdictions, diagnosis is based on a bunch of observations, including lack of response to a painful stimulus, fixed pupils, absence of blinking or other eye reflexes, and electroencephalographic recordings revealing a relative absence of brain activity.

While the general concept is more or less the same across the world, the details can vary, thanks to the fact that the brain doesn't grind to a halt all at once. Loss of permanent function in the outer, 'wrinkled' layer of the brain – the neocortex – sees an end to functions we associate with human consciousness. Any ability to sense, volunteer movement or think about the world is no longer possible. Yet if the nub of tissue at the base called the brain stem is still doing its job, involuntary actions such as breathing and regulating the heart can persist unassisted, and we can hope for life to return to the wrinkled sections of nervous tissue.

Across most of Australia[Δ] and the United States,[ΔΔ] a person is declared legally dead only once the circulation of blood has stopped and there is an absence of activity across the entire brain. By comparison, in the United Kingdom there is no legal definition as such, though the guidelines for confirming a person is deceased state that it's enough for just the brain stem to be silent.[20]

In the past, it was often impossible to tell the difference between a stopped heart and one so feeble that it was eluding our best efforts to detect it, especially when we had little more than our fingertips to feel for a pulse. The premise of a *wake*, where the body is watched for a while by friends and relatives, is thought to have its origins in the fear of being declared dead prematurely.

Horrifying accounts abound of accidental interment being discovered either in the nick of time, or through ghastly evidence in the form of scratches or injury. A pamphlet published in 1661 called *The Most Lamentable and Deplorable Accident* told of mourners in a nearby chapel hearing shrieks from the tomb of a Mr Lawrence Cawthorn; on digging him up, it was discovered, too late, that his shroud was torn and his 'eyes were swollen and the brains beaten

---

[Δ]  All states excluding Western Australia, which has no official position. (See Tobin, B. (1997) Certifying Death: The Brain Function Criterion. National Health and Medical Research Council, Discussion Paper no. 4.)

[ΔΔ]  Most US states have modelled their laws on the *Uniform Determination of Death Act* (UDDA).

out of the head, and clots of blood were to be seen at the mouth, and the breast all black and blue'.

Advances in technology and better medical guidelines mean it's rare for death to be declared accidentally today. Not that it doesn't happen. Rare as it is, the occasional mistaken declaration still occurs from time to time. In Durban, South Africa, in 2016, a 28-year-old male motorcyclist named Msizi Mkhize was declared dead after doctors attempted to revive him following an accident, only to have morgue workers discover he was still breathing when his family arrived the next morning.[21] Similarly, in 2018 three separate doctors gave a cataleptic prisoner named Gonzalo Montoya Jiménez up for death, and he was even marked up for an autopsy. Only when the 29-year-old made a snoring noise inside his body bag did morgue attendants realise the mistake.[22] Every few years a story pops up in the tabloids of some poor soul waking in the fridge and scaring the life out of hospital staff, showing how fuzzy that distinction between life and death can be.

Our trepidation over the very possibility that a mistake could be made in confirming death has a rather unfortunate effect: it can influence our decisions over whether to donate our own or our loved one's organs.[23] Many of us sense that the possibility of donating a healthy organ might mean doctors won't do all they can to save the patient's life. Research has found that while the public often supports organ donation in principle, when it comes to signing on the dotted line a distinct lack of trust in the criteria of brain death as 'the end' – especially if the heart is still pumping – means family members often back out.[24] Far from a lack of compassion, to some degree at least a mix of mistrust and misunderstanding lies behind the shortage of donated organs around the world.

It's easy to see why many of us might hesitate in the emotional final moments of a parent's or child's life; at least a fifth of patients who are declared brain dead while on life support continue to retain some control of water and electrolyte balance. There can still be brainwaves present, and even the appearance of pain responses to stimuli. Faced with the decision of saying goodbye while the memories of our partners or parents or children are still fresh, death no longer seems so textbook. When it feels as if a family member could still wake at any moment, it takes a strong heart to make such a hard, if charitable decision.

In countries where citizens must opt out of having their organs donated on death – such as France, Austria, Belgium and the Netherlands – organ donations can double or more. In the case of Singapore, kidney donations jumped from 4.7 per million citizens to 31.3 per million following the introduction of an opt-out donation scheme.[25] In the warm light of day, the line between life and death seems so sharp that as many as 80 per cent to 90 per cent of us would happily give up a heart or kidney after we've died, and wouldn't dream of going through the effort of opting out of a state-run donation program. After all, we won't need it where we're going, right?

Those good intentions don't always translate into the effort of signing up for a donation scheme, with only around one-third to two-thirds of people making the pledge official by opting in to a donation plan. Mistrust is just part of the reason; a lot of us don't like to bring it up in conversation with family members. Asking a partner at the dinner table if they're going to want their kidney – as if it's the last piece of apple pie – feels a little dark; it always seems like a chat that can wait until tomorrow.

So find a bookmark. Go on. The rest of this chapter can wait. A patient in dire need of an organ, on the other hand ...

If giving up a cornea or two seems no problem, how would you feel about having your entire body preserved after your death and used for educational purposes? Personally, I'm all for it. I'm not sure I would be so confident about it if I'd been born a couple of centuries ago, though.

Studying anatomy was a little less fun when you had to compete against the clock to dissect a body. In the latter half of the eighteenth century, a surge in the number of anatomy schools across Europe didn't help matters, especially when it was already exceedingly rare for anybody to agree to be put on a dissection table after their death. Some tutors made ludicrous promises of a body per student in every course, a marketing technique that probably boosted their enrolments but would have led to a lot of disappointed graduates.

It wasn't just students: even well-seasoned doctors had a hard time getting their hands on bodies to analyse or develop their techniques. In the United Kingdom, the *Murder Act* of 1752 eventually offered up the bodies of executed criminals, but even that seemingly bottomless source wasn't enough to meet demand. Macabre scenes of squabbles between doctors and family members over the still warm, sometimes still dangling bodies of the convicted reflect the desperation of teachers and researchers for subjects. Eventually this prompted some to turn to other, even more unsavoury sources of fresh meat.

Grave robbery was by no means a new crime, but by the dawn of the nineteenth century thieves were lifting more than valuables from exhumed caskets – they were taking the occupants as well. Those who could afford fancy locked tombs or even guards to stand watch might gain a small sense of security, though the truth was that nobody was safe from the so-called resurrectionists.

Echoes of this mistrust between anatomists and the grieving can be found in Mary Shelley's *Frankenstein*, which, being first published in 1818, came out at the height of the body-snatching mania. In contrast to later renditions of the story, the public in the novel would have sympathised with the 'monster' who was stitched together from stolen body parts and brought back to life by an ambitious, overconfident anatomist. The twentieth century and its penchant for horror films might have cast the creation as something to fear, but at the time of Shelley the anatomist was the obvious enemy.

One particularly sympathetic case was that of Charles Byrne, a famously tall man known as 'the Irish Giant'. His greatest fear wasn't death, but that an anatomist and surgeon named John Hunter would get his hands on his 2.3-metre body. So he had it arranged that on his passing, which came in 1783, his friends would put him into a lead coffin and cast him into the sea. Before they could accomplish the deed, a band of body-snatchers paid by Hunter intercepted the pallbearers, secretly swapped the body for a pile of stones and made off with the giant's remains.

I've seen Byrne's bones on display in Hunter's museum at the Royal College of Surgeons in London, sadly reflecting on his fate. Shades of Mungo Man hang about his remains – a tension between the respectful retirement of his body and its continued display for some greater good: attracting income for the museum, to provide future researchers with anatomical knowledge perhaps, or maybe just in the name of educating the general public in the history of the famous surgeon who owned them.

So prolific was the demand for fresh cadavers in the nineteenth century – some going so far as to commit murder just to take advantage of the rich rewards from an unquestioning anatomist – that the British government finally gave in to the demands of petitioners and passed the *Anatomy Act* of 1832. Among its numerous provisions, which included the need for anatomy teachers to be licensed by the Home Secretary, was a section stating that any cadaver in a licence holder's possession could be dissected if (and only if) none of their relatives opposed it. Like today's organ donation, it turned an opt-in system into an opt-out one, making silent or absent relatives acquiescent ones.

The act solved the problem of the resurrectionists, though as effective and fair as it might sound, the new laws opened the way for orphanages and workhouses – consisting largely of lonely women and children – to hand over their dead without question. The poor could take advantage of free burials, once the students had had their way.

Either way, the homeless, estranged and impoverished were subsequently left with fewer rights over their own bodies than the well-to-do and well-connected, a loss that gave rise to years of protest. Not that the voices made much of a difference – the *Anatomy Act* wasn't officially replaced in England, Northern Ireland and Wales until 2004, replaced by the *Human Tissue Act*. It remains in effect in Scotland.

Modern preservation techniques, refrigeration and non-invasive imaging technology mean a cadaver can go a lot further as a study aid than it would have a century or two ago. Still, don't think for a second that bodies are no longer traded on a cadaver market. Brokers for human body parts do exist – there are dozens in the United States, often unchecked and unregulated, offering researchers or technicians limbs, heads or torsos for hundreds to thousands of dollars a pop.[26]

If that makes you squeamish, it might help to know that such funds can help pay for funeral costs in families who couldn't otherwise afford them. On the other hand, such a reminder is reminiscent of the post–*Anatomy Act* economic divide.

For as long as we live and breathe, we have a degree of autonomy over our bodies – if solely by virtue of our ability to speak up about our needs. As with getting sick, the process of dying comes with a relaxing of those rules. Our agency diminishes and our bodies become the property and responsibility of the community, to divide, to study, to dispose of.

If the question of who owns your meaty bits when you're no longer around to argue poses a moral dilemma, the problem becomes exponentially more challenging when we ask who owns our health when we're still alive and well. For the most part, it's easy for me to say, as I have above, that our bodies, and therefore our lives, are thought to belong to ourselves. But that isn't always true.

# SUICIDAL

At the base of Mount Fuji there's a vast forest named Aokigahara, a word that translates roughly into 'Green Sea of Trees'. More commonly it's called Kuroi Jukai – 'the black forest'. It has a reputation for stunning beauty, volcanic geology and a dark history. The woodland grows on a lava field, which explains the rough hollows and numerous sinkholes peppering the uneven terrain that rolls like lithified waves.

The silence deep inside the wood is disquieting, with the occasional snare-drum crash in the distance as a branch or footstep connects with the leaf litter. If there's a wind, it rarely pushes its way down to the forest floor. Maybe it's that absence of noise, or perhaps the haunting beauty, that has made Kuroi Jukai a favourite spot for people to die.

Each year scores of people enter the shadows with the intention of never leaving. There is a sign at one of the entrances, stating, 'Your life is something precious that was given to you by your parents,' urging anybody thinking of taking their own life to reconsider. There are searches for those who ignore the advice, the yellow tape marking an annual coordinated effort to clear the forest of people who made the decision to stop living. Not all of the lost are found in the dense, sprawling ocean of trees.

In 2003 a confirmed total of 105 people crossed Aokigahara's boundary to die by suicide, up from the previous year's tally by twenty-seven.[27] It hasn't always been this way. Historically, the forest's darkest secret was its occasional use by local communities as a site for the practice of *ubasute* – a form of euthanasia, where the infirm, especially the extremely old, were left to starve in the cool serenity of its branches. It took a popular 1960s novel called *Tower of Waves* by Matsumoto Seichō to romanticise Jukai as a spot to see out your last days, its depiction of a lover's death in the forest written with poetic beauty.

I was surprised to learn when I visited Japan in 2010 that suicide was still a polarising topic in such a proud, rich nation. A delayed train excused with a euphemism – *jinshin jiko*, translated as 'human accident' – hinted at something darker one morning. These events are so common, I was told by my host, that Japan Rail reportedly has a policy of charging the 'offender's' family 150 million yen (roughly US$1.3 million) for reparations. Not all families can pay, but the threat of leaving behind such a debt is thought to serve as a disincentive.

The nation's suicide statistics are high enough to make it the leading cause of death for Japanese men between twenty and forty-four years of age, and for women between fifteen and thirty-four.[28] At its peak, in 1958, there were 25.7 suicides for every 100,000 citizens. In the early 2000s the nation saw a small spike in suicide rates in line with those of other nations, precipitated by the global economic downturn and unemployment. Numbers have since fallen again thanks to a strengthening economy and government public health initiatives, but at 15.4 suicides per 100,000 people in 2015, Japan still has proportionately more people who take their own lives than Australia and Canada (both of which are at about 10.4 per 100,000) and the United States (about 12.6 per 100,000).[29]

This isn't to highlight Japan as a world leader in the rate of voluntary death, especially today. In fact, the nation's suicide rates are currently the lowest they've been in over two decades. According to the World Health Organization, at the last official count, in 2015, Sri Lanka was the most suicide-prone nation, with 34.6 per 100,000, a little ahead of Guyana at 30.6. Since official records began, Japan has never been higher than ninth on the ladder.

Where Japan is remarkable is in its cultural relationship with suicide. In many nations across the Western world, suicide is primarily regarded as an act of desperation or mental weakness.

For those in Japan, ending life at one's own discretion has long been associated with stories of honour, love and the preservation of dignity.

Perhaps the most historically renowned tradition is in the samurai's concept of an honourable death through disembowelling, or *seppuku*. In this ancient, deeply respected practice, a captured or disgraced warrior has a chance of redeeming themselves – either of their own volition or by a master's decree – by driving a short blade called a *tantō* deep into their abdomen and slicing across. With courage and some luck, they could sever the abdominal section of the aorta and die quickly from blood loss. By the start of the Edo period, in the seventeenth century, *seppuku* had become greatly ritualised. The process followed strict steps: the samurai wore formal attire, and was surrounded by witnesses and attended by a skilled swordsman, who would cut most of the way through the neck immediately after the first incision. Wives would often follow their husbands in what was called *jigai*, an old word that became fashionable again in the modern age thanks to narratives such as Matsumoto's.

Of more recent notoriety are World War II's *kamikaze* 'divine wind' fighter pilots, who would fly their planes into enemy units. Under the direction of Emperor Shōwa, volunteers were selected for special units to fly cheap aircraft with old engines on missions with the purpose of crashing them into the enemy as weapons. The first *kamikaze* attack killed thirty Australians on a navy ship. By the end of the war, in spite of only about 10 per cent to 20 per cent of attacks hitting their target, several thousand pilots had sacrificed themselves, taking around 5000 enemy lives in the process.[30]

As a military strategy, *kamikaze* pilots were arguably pretty ineffective. In the West, their devotion was used to paint the Japanese as unreasonable. In propaganda, portraying the pilots as suicidal was a convenient way to represent the enemy as mindless, if not downright insane.

Post-war twentieth-century Japan was a time of rebuilding, not just economically but also culturally. Social commentators patriotically referred to their country as a *jisatsu no kuni*, or 'a suicide nation'. The reflection within the literature and media wasn't entirely pathological, either. There was almost a sense of reluctant pride in the admission, a feature of the national character inherited from a tradition that wasn't pretty, but was nonetheless honourable. Japanese attitudes towards voluntary death ranged from tolerant to reluctantly expected, though the concept of suicide has remained a nuanced one that can't be understood by reference to numbers alone. If the history of Japanese culture is anything to go by, not all voluntary ends are the same.

'Only Japanese people can understand the suicide of the Japanese,' author Ōhara Kenshirō wrote in his 1965 book *Nihon no Jisatsu* ('Suicide of Japan'). 'Foreign scholars can look at the statistical numbers on suicide, but they will not understand the phenomenon,' he concluded.

Putting the numbers aside for the moment, suicide in Japan is an issue in constant evolution, shaped by centuries of Japanese cultural history and Western medical influence. In some ways the recent embrace of suicide as a strong part of the national character has represented a rebellion against outside intrusion. Officially, Japan has worked hard to drive down suicide numbers with studies and public health programs that aim to intervene and provide help. Unofficially, though, there remains a sentiment that resonates throughout the media, being subtly evident in the press, manga, novels, soaps and movies.

There isn't a clean split between Western and Eastern values with respect to suicide. How many of us look back on Socrates taking hemlock after being found guilty of impiety by corrupting the youth of Athens as a form of heroism, sacrificing himself in the name of political progress? We still revere Shakespeare's play in which his famous star-crossed teenage lovers throw the ultimate tantrum by drinking poison, an act of romantic resistance rather than insanity or weakness. Western history distinguishes between respectful and pitiful intentions when it comes to us taking our own lives just as definitively as Japan's.

Indeed, most cultures celebrate last stands and military sacrifices not as selfish or cowardly acts but as deeds of respected defiance that serve a greater good. Religions abound with martyrs who gave up their life in the name of faith rather than renouncing or hiding their beliefs. Christianity is based around a deity's knowledge of his impending demise, and doing nothing to prevent it in the hope of redeeming humanity. Islamic suicide bombers, while a minority, have dominated the news in the modern age of terrorism. Buddhists have made the front pages of newspapers by peacefully meeting death through self-immolation in decades gone by. The stories we tell and celebrate all over the globe reflect humanity's complicated relationship with choosing death, so Japan isn't exactly extreme in its self-reflection of what it means to have a say in when we stop living.

That said, there is a strong historical tradition in the Western hemisphere of debating the moral ambiguities of suicide. Theologians such as Thomas Aquinas declared it to be a sin against God, not to mention a breach of natural law and an offence against the community. On the other hand, philosophers including David Hume and Friedrich Nietzsche have argued that it's the ultimate act of individual freedom. Whether – or under what circumstances – it's permissible to choose to die has been a meaty topic for discussion in law and philosophy for centuries.

At one end of the scale are those who feel that the choice of whether to continue living or not is entirely up to the individual. For those at the other end, there are no acceptable circumstances in which one should volunteer their own death. In between is a spectrum of justifications and conditions based on benefits to the community, prognosis of suffering and opportunity for recovery. In the form of assisted dying and euthanasia, suicide takes on a new context – it's the ultimate cure, but comes with the most costly of all side-effects.

The laws on euthanasia and assisted dying are inconsistent across the globe. Physicians in Colombia, the Netherlands, Luxembourg and Belgium can't be prosecuted if they administer life-ending treatments to patients who request them. In Switzerland, Japan, Germany, Canada and half a dozen US states, including California and Oregon, patients can be prescribed similar treatments which they must self-administer in what's referred to as physician-assisted dying. Australia's Northern Territory briefly permitted both direct and assisted euthanasia in 1995, before the *Rights of the Terminally Ill Act* was nullified in 1997. A law permitting euthanasia was passed in Victoria in 2017, and is expected to come into force in 2019. Euthanasia and assisted dying remain hotly debated topics across the country.

And they are not trivial matters: they force us to choose between the suffering of individuals and the suffering of a community; the right to govern our own bodies and the right for society to preserve its members; the value of an individual's dignity and the abstract notion of human existence.

As with organ donation, there is the question of trust. We all like to think we'd prefer a peaceful death over prolonged agony and a loss of independence, but when reality hits, how many trust the system to act accordingly and responsibly, without undue haste or bias? Then there is the question of transience of value. Suicide is an infinite solution to a finite experience. No matter how much pain we're in or how much indignity we're suffering, there's the possibility that our situation could change in the future. Time makes a new person of all of us, and it's valid to question who of us has the right to deprive a future version of our self of the right to live.

Lastly, we can ask why we should distinguish one form of suffering from another. Why not provide prisoners on death row with the same swift access to death, rather than endure years in relative isolation as court appeals and bureaucracy eat away at their sanity? What of adolescents with permanent debilitating conditions that are not life-threatening? Following the trajectory of that 'slippery slope', what of those who simply no longer value life at all? Why not permit anybody who lacks happiness to die?

Anhedonia is considered a symptom of conditions such as depression and schizophrenia, though isn't an illness on its own any more than pain is. It's characterised by an emotional numbness, a life without joy. In 2015 *The Economist* published a short documentary featuring Emily, a 24-year-old woman from Belgium who expressed a will to make use of her nation's relatively relaxed euthanasia laws to end her life. In her words, she'd had enough. 'I feel like nothing gets to me anymore, like I'm dead inside,' she claimed.[31]

The process takes at least two years and requires three doctors to sign off on the applicant's prognosis: each of them must determine whether the patient's state is likely to improve in the future. Mental suffering is a legitimate criterion for state-sanctioned euthanasia in Belgium, yet it's hard to draw the line between a life without happiness and one in active discomfort – if there is a distinction at all.

Emily was ultimately granted permission, though at the time of writing she's yet to go through with the process. Some regard this as a sign of the law's weakness, demonstrating an inability to confirm beyond doubt that suffering is ever truly endless; others think being permitted to die is a freedom that – ironically – buys relief. Many, like Emily, find life's pains easier to cope with once they know an assisted death is within reach.

A perplexing paradox at the heart of euthanasia is the duality of disease. On one hand, disease gives rise to mental suffering. On the other, it robs us of agency. To what degree does the suffering produced by mental illness – be it bipolar disorder, schizophrenia or chronic depression – justify ending one's life? And to what degree does it rob us of the choice?

The desire to die has evolved into a hallmark of mental illness, as a consequence of centuries of debate dominated by medical science. Suicide as a symptom of disease has its roots back in the writings of an early-nineteenth-century French scholar and psychiatric pioneer by the name of Jean-Étienne Dominique Esquirol.

In 1818 Esquirol completed a three-year tour of France's asylums. He came back less than satisfied, as he'd found that the so-called insane were treated little better (and often worse) than criminals. Something had to change. By introducing new methods for analysing, describing and classifying mental health, Esquirol helped revolutionise the widespread treatment of patients. It's thanks to him we have the word *hallucination*. Yet it's also largely thanks to Esquirol that we see suicide as the product of a broken mind.

His 1838 publication *Mental Maladies: A Treatise on Insanity* was the first clinical attempt to categorise disorders of the mind by attempting to pull apart the consequences of physical damage from developmental problems that seemed to have no clear cause. He distinguished mental development from mental illness, drawing a line between disabilities and diseases of the mind. Attempts at suicide that were previously seen as immoral and something to judge, according to Esquirol, were instead pathological signs of impaired decision-making. Across much of the world, suicide had legally been – and in many cases remains – an act akin to murder. Esquirol didn't see it that way.

Killing oneself was established as a common law crime in England in the thirteenth century, reflecting a long history of regarding it as a grave sin. *Felo de se* – or 'felon of oneself' – resulted in being denied a hallowed burial, and the family being denied their inheritance, especially if it was connected to a crime or a desire to cause harm to somebody else.

An English cleric named Henry de Bracton wrote on the philosophy and letter of the laws of his day in the famous treatise *De Legibus et Consuetudinibus Angliae* ('On the Laws and Customs of England'):

> [I]f one lays violent hands upon himself without justification, through anger and ill-will, as where wishing to injure another but unable to accomplish his intention he kills himself, he is to be punished and shall have no successor, because the felony he intended to commit against the other is proved and punished, for one who does not spare himself would hardly have spared others, had he the power.[32]

The person's mental state was often taken into consideration, with madness removing some, but not all, culpability. As de Bracton noted, 'But if a man slays himself in weariness of life or because he is unwilling to endure further bodily pain ... he may have a successor, but his moveable goods are confiscated.'

Similar laws abounded across Europe throughout the Middle Ages, with the bodies of those who had taken their own lives being buried or burned, in contrast to a respectful religious ritual. Their possessions were also confiscated by the state. In Germany, the bodies of those who died by their own hand were to be buried beneath the gallows. In France, their remains could be legally drawn and quartered.

Change was gradual, but philosophical debates challenged the theological doctrines that damned those who took their own lives, leading to greater secular influence over the law. Shakespeare's plays reflect this shift in discussion, famously framed in Hamlet's 'To be, or not to be' soliloquy. The shift in literature with the mid-eighteenth-century Romantic poets and their sentimental representations of suicide reflected a similar trend in modern Japan.

The focus turned from suicide as a thing that we morally decided, to something that was pathological. And Esquirol was on the front line of the debate. 'It does not belong to my subject to treat of suicide in its legal relations, nor, consequently, of its criminality,' the French psychiatrist wrote. 'I must limit myself to showing it to be one of the most important subjects of clinical medicine.' He went on to acknowledge that suicide is a vastly more complex phenomenon than was usually assumed, and one where there were diverse causes and contexts. The ones that interested him most were the passions of the mind amid their 'seasons of fury'. 'In their excesses, there is nothing that they do not sacrifice,' he wrote, 'and man, while a prey to a passion, spares not his own life.'

What began with Esquirol developed into a perspective on suicide that turned it from an act of pure choice into a neurological flaw. Again, the shift was gradual: England and Wales took until 1961 to wipe suicide from the common law books.[33] Only five years earlier, of 5387 failed suicide attempts known to police, 613 people were prosecuted; of those, thirty-three spent time in prison.[34]

It's still illegal to attempt suicide in dozens of countries around the world, from India to Singapore, Malaysia and Saudi Arabia, and in most places it's punishable with a prison term. Killing yourself in North Korea will see your family punished. Most states or provinces in the United States, Australia and Canada no longer regard suicide as a crime, though some – such as the US state of Virginia[35] – remain as exceptions.

Suicide in the West is much more often assumed to be the product of a broken mind than of an immoral one. Some psychiatrists go so far to make the bold claim that 'a psychiatric disorder is a necessary condition for suicide to occur'.[36] Even when individuals are diagnosed as 'psychiatrically normal', researchers will occasionally speculate that there might be some 'underlying psychiatric process that the psychological autopsy method, as commonly carried out, failed to detect'.[37]

Confusingly – or perhaps tellingly – statistics across diverse cultures tend to paint dramatically different pictures. Studies on rural Chinese people aged fifteen to thirty-four, for example, have found that signs and symptoms of acute mental health disorders are identifiable in fewer than half of all suicides.[38] In Japan, the figure is closer to two-thirds.[39]

Another piece of research on Chinese youth concluded that a diagnosis of mental illness 'was an important predictor of suicide in males, but not in females'.[40] A study on young people from rural India found that just over a third of those who took their own lives had a *DSM-III-R* psychiatric diagnosis.[41] An analysis of Malay medical records in 2014 arrived at an overall figure of just over one in five.[42] A 2004 evaluation that combined twenty-seven separate studies, mostly covering Western populations, concluded that 87.3 per cent of those who killed themselves had a mental illness, but noted that the figures ranged from 63 per cent for a Chinese study to 98 per cent for Taiwanese, Scottish and Finnish research papers.[43]

That's a lot of different numbers – what could explain such variety? It seems as if there's hardly a number left untouched, given the huge number of studies, from 22 per cent of suicides being associated on some level with poor health, to virtually all being the direct consequence of mental illness.

In the words of one set of researchers, it's important to consider the 'social and cultural factors influencing how one views and interprets suicide, and cultural biases towards or against specific diagnoses'.[44] Even before we begin to collect data, a study makes some basic assumptions about what constitutes a mental illness. Different methodologies also make it hard to compare research, with some conclusions based largely on coroners' reports, or deaths occurring in areas where mental illness simply went undiagnosed.

Psychological autopsies – the post-mortem equivalent of a clinical psychological assessment – rely on interviewing friends and family and analysing the deceased's personal communications. These methods aren't standardised, and have been recognised as being notoriously open to intrusive cultural biases. An American doctor's impression of clinical depression could look just like an overworked banker to a Chinese physician; a Japanese psychologist could diagnose a patient with *hikikomori* – a word that means 'to pull in', describing a condition of social withdrawal among adolescents and adults – when in Britain they might be dismissed as just another sulky teenager.

What this all adds up to is a big question mark over the relationship between suicide and mental illness. There's little doubt that our neurological wiring – especially that which correlates with conditions responsible for schizophrenia, depression, addiction and personality disorders – makes it more likely that a person might consider and value suicide. Even if the neurology isn't always directly to blame, for many people the dysphoria that comes with the struggle to cope with mental illness in a world that sees them as bad or broken can also make suicide seem like an appealing solution.

It's not unfair to say mental illness raises the risk of suicide. It's an absurd leap of logic, though, to conclude that a desire to die constitutes a pathology all of its own.

The balance might be informed by biology, but the complete answer won't be found in the sequences of our genes or the levels of our neurotransmitters. Appealing to authorities to make the decision for us, whether gods or lawmakers, only serves to make saints of some and sinners of others.

No doubt we'll be debating suicide for a long time to come, but if anything is clear, it's that the desire to die is far more complex than the simple dichotomy of evil or pathological. If we're to fairly decide when a person has a right to choose their own fate and when an authority should intervene, we must move on from asking whether suicide is a sickness, and ask instead what value we place on ending an individual's suffering.

# HUMAN

Not one to brag, but I've seen a lot of sperm in my lifetime.

One of my least favourite tasks at the lab was to open freshly delivered jars of body-warm semen and count the number of squirming cells down a microscope lens. I have fairly decent eyesight, so believe me when I tell you that not a single sperm had a minuscule baby inside it.

Microscopes haven't always been so powerful, so you can forgive the world's first microbiologists for believing the sperm cells they looked at did hold an itty-bitty human body. The seventeenth-century Dutch tradesman and amateur natural philosopher Antonie van Leeuwenhoek did what nearly any curious person would do when he got his hands on one of the world's first quality compound microscopes – he looked at whatever he could get his hands on. This included semen. (Because of course he did.) And so when he and a student (that must have been one hell of a conversation) named Nicolaas Hartsoeker looked at the hazy, writhing filaments, a mix of refracted light and healthy imagination meant they thought they could make out a tiny head and limbs inside the cell.

In a time long before we knew that sperm carried the father's contribution of genetic material, it was believed that something inside semen contained an expandable baby. Like a seed in a field, the baby was planted into a womb, which provided the necessary liquids and nourishment to make it bigger. One proponent of this 'preformationist' hypothesis, Nicolas Malebranche, speculated that all of Earth's life forms were created at the same time many generations ago, and tucked away inside their ancestors like the world's longest line of matryoshka dolls.

So when Hartsoeker looked at individual sperm cells, he expected to see something more human-like, inspiring him to draw the first depictions of human spermatozoa containing a bundled-up *'petite l'infant'*. His work set the standards for embryology for a long time to come, only changing as improvements in microscopes eventually demonstrated that there were in fact no tiny people inside sperm cells.[45]

The contrasting hypothesis, called *epigenesis*, claims that individual organisms form out of simpler components arranged in complex ways. By the 1780s, both flavours of preformationism – the one that saw the sperm as the deliverer of the future human, and the other that blamed the egg – had fallen out of favour. Thanks to microscopy and studies of things like frogs and sea urchins, it was widely accepted that an animal took shape as individual cells changed, split and multiplied.

This raises an interesting question: when does a bubble of fat, DNA and protein become a human being? We might as well ask how many grains of rice make a pudding, or how many hairs must we lose to become bald. Somewhere between a single sperm chewing its way through the membrane of an egg cell and a newborn's first breath, we can say that tissue becomes a bona fide human.

As with death, biology can paint us a detailed picture. It tells us that only around half of all of the eggs that are successfully fertilised will successfully bury themselves into the lining of the uterine wall. And of those that do, between a third and a half will be miscarried, typically long before there's any sign of pregnancy.[46]

Biology tells us that just a few weeks after conception, tissues that will eventually become a heart will start to flutter in a pulse-like fashion. Shortly after, a mere four to five weeks after fertilisation, a simple layer of tissue folds up inside a tube to form the very beginnings of a brain. By week seven, the first nerve endings are connecting into a network we can recognise as the foundations of a peripheral nervous system. Before the end of the second month, the inch-sized blob of tissues has physical characteristics that look almost human-like: a large head, dark spots for eyes, tiny buds for hands and feet.

Biology also tells us that a foetus can respond to sounds at around sixteen weeks, before the ear has completely developed. It tells us that there is an outside chance a foetus delivered into the world at twenty-one weeks can survive, if it gets medical assistance. Wait another three weeks and those odds become an even fifty-fifty.

By twenty-six weeks the foetus shows responses to noxious stimuli. It can open its eyes around the twenty-seventh week, with the rapid eye movement indicative of dreaming becoming obvious a short time later. What's more, biology tells us that birth isn't the end of development, with the brain continuing to grow and make connections for years to come.

What science is incapable of doing is dictating which – if any – of those characteristics are required for a developing human to be granted rights and a moral status. Biology merely sets the scene in which a human animal emerges; it's up to us to debate the moment when a human becomes a person. Because of this, the line separating one generation from the next has shifted back and forth down the ages, from the moment of conception to sometime after the birth.

There is evidence that humans have intentionally ended their pregnancies prematurely since prehistoric times. For as long as we've recorded medical treatments, we have listed medications and procedures that have acted as contraceptives and abortives as casually as if an unwanted pregnancy were akin to a kidney stone or stomach ache.

Galen advised barrenwort and juniper, among other herbs and poultices. Pliny the Elder offered the warning (or was it advice?) that stepping over a live viper could cause miscarriage – unless the woman happened to have a two-headed serpent in a box with her (an essential item for any modern woman on the go).[47] Writings attributed to Hippocrates advised sex workers to jump up and down to 'loosen the seed' if they didn't want to fall pregnant – even though in his famous oath the physician also claimed: 'I will neither give a deadly drug to anybody who asked for it, nor will I make a suggestion to this effect. Similarly I will not give to a woman an abortive remedy.'

The debate over the moral and legal rights to terminate a pregnancy have mostly revolved around the question of when a developing embryo should be thought of as a person. Science, as we've seen, has no clear position on this. Most religious concepts of personhood, on the other hand, are about embodiment – the delivery of what we might think of as a spirit.

Odd as it might seem, the Judeo-Christian scriptures have little to say on precisely when a prenatal life becomes a person, leaving those of faith to dig deeper for hints. Theologians have debated the relevance of Exodus 21:22: 'If men strive and hurt a woman with child, so that her fruit departs from her, and yet no mischief follows: he shall be surely punished, according as the woman's husband will lay upon him; and he shall pay as the judges determine.' By some accounts, this seems to imply assault causing a miscarriage, which would result in a fine. Hardly the 'eye for an eye' of murder, or exile for manslaughter. But translations of words in ancient cultures aren't always straightforward. Taken at face value, the precise combination of the Hebrew terms – *yeled* for 'child', and *yasa* for 'going' or 'delivering' – seems to describe a child being born.[48] Nowhere else in the scriptures making up the Old Testament do the words refer to death or a stillbirth, so it could be argued that an assault that leads to a *premature birth* is a crime, but not necessarily murder.

Psalms 51:5 is also taken as a clue to when a soul might enter a body. Commonly translated to read, 'Behold, I was shapen in iniquity; and in sin did my mother conceive me,' it could be reasoned that a soul had to be present on conception in order for it to be sinful. Unfortunately, we're again hit by the difficulties in translating words that are no longer in common use. The original word *yechemathni*, which is often assumed to mean 'conception', refers to the act of nourishment or warming; it could also broadly mean 'gestation' in its entirety.[49]

Nonetheless, medieval Christian philosophers gleaned what they could from Greek translations of such old texts, combining them with the classical views of Plato and Aristotle, to come to a range of conclusions. St Jerome and St Augustine claimed that an infant was a person only when it started to look like one, which could mean from the second trimester.

St Thomas Aquinas was heavily influenced by Aristotle's argument, so in his work *Summa Theologica* he sided with the ancient philosopher's view that male embryos inherited their rational spirit a full forty days after conception, while female ones didn't get theirs until around the eighty-day mark. This timing is roughly equivalent to what has been termed the 'quickening', referring to the first time a mother feels her unborn child moving. Given this generous amount of time between conception and personhood, for much of history it has been morally acceptable to terminate a pregnancy so long as there has been no sensation of movement.

By the same token, to be 'quick with child' could also offer a condemned woman a temporary reprieve. 'To be saved from the gallows, a woman must be quick with child – for [to be] barely with child, unless he be alive in the womb, is not sufficient,' remarked an English juror in 1770.[50]

In 1803, English statute law made it a crime punishable by death to medically terminate gestation after the quickening. Less than four decades later, any reference to quickening was abandoned – the act of inducing an abortion would no longer end in a hanging, but the movement of a foetus no longer distinguished it as a distinct human life.

It's likely that scientific advances in the eighteenth and nineteenth centuries would have challenged existing notions about when a bundle of cells is complex enough to be considered a person. New models of epigenesis were becoming increasingly detailed, with progress in anatomical dissections and improved microscope techniques shedding light on how tissues develop in growing embryos. Quickening was an outdated concept once it became clear that complex, human-like features emerge at various stages of gestation.

Today we tend to divide personhood between conception, viability and birth, depending on who's asking. If a foetus makes it to anywhere between twenty to twenty-eight weeks and then presents no signs of life, it is considered a stillbirth. In many jurisdictions the delivery can be legally recognised as a birth, even if a death certificate isn't always granted.

Abortion attitudes and legislations vary across the globe. In most European countries, gestation can be terminated up to fourteen weeks, or up to twenty-four weeks if there are extenuating circumstances, such as a threat to the mother's life. In Australia abortion remains a criminal offence in Queensland and New South Wales, with exceptions granted where the foetus's condition is considered 'inconsistent' with life, or if continuing the pregnancy puts the mother in danger. In states and territories where it is legal, the maximum age of the foetus varies, anywhere from fourteen to twenty-four weeks. In the United States, a judicial interpretation of the Constitution means states aren't permitted to ban the procedure outright, but they can still pass laws that implement restrictions, such as extensive waiting periods, or limit the availability of facilities.[Δ]

The premature termination of a life is where health and body autonomy merge into a messy tangle of values. Nature can't dictate a precise moment when the tissues and organs inside a womb deserve equal rights to those outside of it. It has no authority on which criteria – if any – need to be met before a potential lifetime can be cut short. Those are political decisions, ones that we as a society must agree upon by determining the hierarchy of our priorities through public discussion.

Each and every one of us has a finite lifespan, at least for now. Our ability to comprehend our own expiry date, as well as those of others in our community – extant and potential – is unique among animal brains. Only modern humans and their close ancestors seem to have evolved minds capable of contemplating the moments when life starts and stops. To what extent does this knowledge of our

---

[Δ] The famous *Roe v. Wade* case of 1973 was a landmark decision that said states can't make abortion itself illegal, but even so they don't have to make it easy for women to access.

future non-existence affect our behaviour? A twentieth-century anthropologist by the name of Ernest Becker was confident that death wasn't just an uncomfortable thought – it was fundamentally responsible for a great deal of human culture. Becker argued that our behaviour as thinking animals isn't geared just towards physical survival, but also towards symbolic permanence. In words attributed to the famous street artist Banksy, 'I mean, they say you die twice. One time when you stop breathing and a second time, a bit later on, when somebody says your name for the last time.'[Δ]

Becker suggested that we all want that second death to be as far into the future as we can make it. 'What does it mean to be a self-conscious animal?' he writes. 'The idea is ludicrous, if it is not monstrous. It means to know that one is food for worms.'[51] Some people find death so impossible, he figured, that they look for ways to outlive their physical form. This could mean invoking a belief in the supernatural, or leaving behind a legacy of discovery or philanthropy, or even having children. So powerful is this force, Becker suggests, it's responsible for our desire for sex and our disgust over other people's choices, and even for giving rise to many mental illnesses.

Becker's writings inspired a model of social behaviour called Terror Management Theory, which describes symbolic elements in our culture that protect us from our uncomfortable sense of mortality. Any threat to our perception of individual worth or self-esteem, such as a reminder of our mortality, causes us to reinforce the meaning of our existence in some way. This reinforcement could be beneficial, prompting us to change our health routine or to embrace a YOLO philosophy, but it could also be socially destructive, giving rise to acts of aggression against anybody who threatens our precious values.

Unfortunately, without a bunch of immortals to turn to for advice, it's hard to know to what extent awareness of our impending

---

Δ  Banksy was allegedly quoted as saying these words of wisdom in a 2010 article in *The Sun* titled 'Banksy in His Own Words'. It's hard to verify if the enigmatic artist did utter them, but they make for a nice sentiment anyway.

doom underpins our basic nature. There's no doubt that the sense that there is 'something more' to our living bodies than clay and air has had a significant influence on many endeavours throughout history – not just in our mortuary rituals and spiritual beliefs, but in our historical attempts to understand our bodies as physical systems.

The inevitability of our end is matched only by the inevitability of disease. Benjamin Franklin's timeless words might as well have been: 'Nothing can be said to be certain, except death, taxes and being unwell.' Yet while awareness of our own failing biology is one thing that unites us as a species, it's also one thing we can't always agree on.

# VI
# WELL

IV

# EXTRAORDINARY

Among the framed family photos on a wall in our family room is a black-and-white snapshot of my son when he was still curled up inside his mother's womb. Like any first-time parents, we stared at that ultrasound picture on the monitor, hearts fluttering with anxiety and hope, desperate to be told 'everything looks normal'.

And if those hadn't been the words we heard? If something had have been seriously wrong? Broken? Unusual? Unwell?

I can count my blessings that I don't have a ready answer.

Nearly a decade has passed since that day. Thankfully, not everything was perfectly normal. Not quite, at least. It turned out that our son's brain is a little different to ours. In many ways it's better. He's capable of feats of recall and insight that I, as a science writer, can only dream of. His imagination is wild and his sense of humour is a little weird, sure, but they're also characteristics I deeply envy. Right now, with his whole life ahead of him, my son shines with hope and promise. His abnormalities – on balance – are oddities his family celebrates rather than suffers.

It isn't all glitter and unicorns, though. He struggles to grasp many of the social quirks of that strangest of animals, *Homo sapiens*, a feature that might or might not become a concern in the near future as he navigates his way through the trials of adolescence. Like many who find themselves somewhere on the autism spectrum – not to mention many of us who don't – the peculiarities of human interactions can be irrational, confusing and often confronting.

My son often repeats phrases before settling on one he likes, a behaviour with the delightful name of *echolalia*. The everyday 'pragmatic' language skills we learn to use in order to communicate the subtleties of etiquette and platitudes don't come easy to him, which can be awkwardly (and humorously) clear in his expressions and mannerisms. Even the most obvious of quips might be met with a concerned look, followed by: 'Is that a joke?'

To compensate, we fill his mind with tips and stock phrases to help him deal with the American and Australian etiquettes he encounters as a dual citizen. But I fear what might happen if – in spite of his wicked intelligence and astonishing memory – his difficulty with nuance opens him to being ostracised or, worse, exploited by strangers and con-artists.

Communication challenges are just part of the story. Sensations I dismiss as merely annoying have been deeply uncomfortable for the poor lad. The buzz of a busy crowd or the boom of fireworks can become a terrifying cacophony in his head, one that in the past has induced levels of panic I can never fully appreciate. To maintain a balance of stimuli, he sometimes wears noise-cancelling headphones, or paces around, clapping and waving his hands in overwhelming joy or concern, or to act out the narrative inside his head, actions that fade into normality at home but risk being misinterpreted by those less informed (and more judgemental).

Our son's condition technically sits just inside the borders of the condition historically called autism, but more often referred to now as autism spectrum disorder (ASD). But it's been hard for us to see how *disorder* is the right word. If anything, my life has become more orderly because of it. His desire for structure means he often disciplines himself. He avoids noise and does his best to negotiate confrontation in order to avoid conflict. Tantrums have been rare, even when he was a toddler. I can't imagine what it would be like having a child less imaginative, self-disciplined, logical, witty, affectionate and, well, weird. He's perfect for us.

Of course, I can't generalise from our experience, and neither should you. Countless parents deal with autism's traits as a daily challenge, with combinations and severities of characteristics resulting in behaviours that are emotionally and physically exhausting, if not occasionally threatening. When stimulation turns into self-harm, or when a stray touch sets off a defensive meltdown, or when communication is limited through a complete absence of speech, it's impossible to pass off ASD as a trivial quirk of personality.

It seems almost trite to emphasise the 'S' in ASD. It goes without saying that the autism umbrella covers an incredibly wide variety of characteristics. While there's a thin thread of features connecting all those with a diagnosis, every individual constitutes a unique picture of the disorder.

And at some point on that landscape of shared traits there is a perimeter. It isn't a solid, crisp line determined by a permutation of genes or an extra chromosome. There's no single obvious wonky protein, no virus we can sift out, no missing organ to distinguish those who are in and those who are out. No simple blood test will tell us if an individual has autism. For now at least, it's a boundary defined by careful observations of behaviours and anecdote, drawn and redrawn by debate, informed by a need to balance the risk of stigma with an ability to cope in an ever-changing society. Not to mention the hard realities of budgets and available resources.

So on one side of the line are the likes of my son, who – fortunately – is entitled to assistance and subsidies. Close by, just on the other side of checklists and expert opinion, there are those who are left to wonder if their slightly atypical personalities have a label, or if they just find certain kinds of socialising a little harder than most. A diagnosis, for us, has been a key to unlock opportunities to adapt – something most of us struggle to do with brains that don't have convenient descriptions and umbrella terms.

For many, an introduction to the autism spectrum comes through popular storytelling: maybe the awkward charm of the surgeon Shaun Murphy or the forensic scientist Temperance Brennan on television shows such as *The Good Doctor* and *Bones*, or the childlike simplicity of Raymond Babbitt in the movie *Rain Man*. So maybe we can be forgiven for buying into the stereotype of the autism spectrum as being full of aloof geniuses who can count fallen toothpicks in a blink but can't fathom the significance of their talent, or academics who solve mysteries with perfect logic but can't grasp why a person might lie in order to cover their shame.

It's a cliché, sure, but the stereotype is ripe for fiction. The conflict between superior intelligence and what seems like an absence of compassion is a slightly softer version of the 'mad genius' trope: it's Faust seeking knowledge at the cost of his soul, or Frankenstein discovering the spark of life but failing to consider the cost of suffering. Not that I'm implying Johann and Victor were on the autism spectrum, but it appears that we've been telling stories for centuries about individuals with godlike knowledge tempered by social naivety, fearful of any ability to contemplate nature unhindered by care and morality.

Stepping back for a moment, the cliché isn't completely without credence. Forget the notion that autism robs individuals of empathy – I promise you, it doesn't, even if negotiating the subtleties of emotion can be a little hard sometimes. And, truth be told, learning difficulties are more common among individuals who have an ASD diagnosis, with roughly half presenting an IQ of less than 70.[1] But what's ironically called 'savant syndrome' does disproportionately (even if not exclusively) affect individuals on the autism spectrum, with as many as one in ten developing extraordinary skills in a narrow range of disciplines, including mathematics, music and art.[2]

We've been fascinated by such contrasts in human mental ability for centuries. The first accounts of such extraordinary talents among individuals considered to be otherwise intellectually delayed date back to the late eighteenth century, with the likes of Jedediah Buxton, who could crunch lengthy exponential equations in his head. Then there was Thomas 'the Virginia Calculator' Fuller, an African-American man enslaved at age fourteen 'who could comprehend scarcely anything, either theoretical or practical, more complex than counting'.[3] He once demonstrated his abilities by mentally calculating the number of seconds in the life of a person who lived to seventy years, seventeen days and twelve hours.[4]

It wasn't until 1887 that physicians made a solid attempt to define this kind of human curiosity as a medical condition. John

Langdon Down – a doctor whose name is forever linked with the chromosomal syndrome trisomy 21 – delivered a lecture based on his life's work as medical superintendent at the Earlswood Asylum in Surrey. Or, to give its full name, the Royal Earlswood Asylum for Idiots. That's right, a word that has come to be a slur for irrational or counterintuitive behaviour was once a legitimate medical label, generally describing people with an IQ estimated to be 25 or less. Down specifically referred to those in his charge with extraordinary mental talents as *idiot savants* – a clinical description that was fashionable for half a century, before it joined other outdated words such as *moron, cretin* and *spastic*.

In the same lecture, Down also discussed young individuals with what he called 'developmental retardation': children who seemed to regress in cognitive skills, despite having apparently developed normally for their first couple of years. These children, he said, presented 'rhythmical and automatic movements' and 'lessened responsiveness to all endearments of friends', appearing to live in a world of their own.

While this sounds an awful lot like autism, the term wouldn't be applied until the 1940s, when an Austrian-American physician named Leo Kanner borrowed a word used to describe the withdrawal of patients diagnosed with schizophrenia. Like Down, Kanner also noticed that the subjects in his case studies appeared to have impressive memories.

Shortly after, a Viennese paediatrician named Hans Asperger described children with what he saw as a highly functional form of autism – a condition that would be linked with his name for half a century, until eventually it was included under the banner of the autism spectrum. Even today there is a tension that promises to tear Asperger's syndrome free from ASD, with some arguing that it should be made a separate diagnosis once again.

To Kanner, these children appeared to be socially detached, withdrawing into their own imaginations and avoiding the confusing mess of stimuli in the busy outside world. Although they seemed to ignore the world, their minds were far from empty. It was as if they'd exchanged normal brain function in one area for improvements in another, sacrificing the complex processes necessary for engaging with other humans in order to devote their mental powers to building connections between numbers, harmonies or geometries.

Kanner initially blamed the parents – especially the mothers – for putting the brakes on development, accusing them of being too 'cold' and detached. A *Time* magazine article quoted him as saying the parents of autistic children were barely able to 'defrost enough to produce a child', an analogy that gave rise to what was famously called the 'refrigerator mother theory' of autism development.[5] Although Kanner later recanted, a controversial figure from Chicago named Bruno Bettelheim reheated the hypothesis of emotionally distant parents, effectively promoting it in the public eye despite having faked his credentials and fabricated his data. It's taken decades for the ugly 'refrigerator parent' model to fade away, and the tendency for parents to feel guilt remains.

The pseudoscience of the British gastroenterologist Andrew Wakefield over the past two decades hasn't been easy for society to shake off either. His 1998 paper linking the measles, mumps and rubella vaccine with autism has done considerable damage, not only for baselessly instilling a sense of regret and shame in many poor parents seeking to do the right thing for their children and society, but also for inspiring a generation of new mothers and fathers to forgo vaccination, out of fear that their kids might inadvertently fall into the autism spectrum as a result.

Even with the knowledge that my son's atypical neurology wasn't caused by his vaccinations, and that our love and attention wasn't deficient, my wife and I cannot help but reflect back on his gestation and early childhood and wonder, 'What happened?'

It almost goes without saying that the reality of ASD is complicated. A useful adage might be: 'If you've met one person with autism, you've met just one person with autism.' Probabilities aside, a gift for mathematics or music isn't a certainty, nor is a fate of ostracism. In my teaching career I've had half a dozen or so students who have been ascertained to be sitting somewhere on the upper end of the spectrum. Not that I can say my classroom career has handed me a solid sample size, but each of these students was unique in their talent and interest in science, art and maths, not to mention in their particular academic needs and their aptitude for making friends and engaging in romantic relationships. Pinpointing any shared biology of people diagnosed with ASD is doubtless going to be a challenge for researchers.

So, if unloving parents and vaccines don't cause autism, what does?

Clearly there are neurological differences involving filtering and translating the constant wash of stimuli we sense, not to mention cues of language and non-verbal communication we generally take as social markers. Decades of brain scans, autopsies and comparisons with mice bred to contain human genes have led back to a range of differences at gross anatomical, connective and cellular levels, implying a broad physiology behind autism's characteristics.

Take, for example, one recent area of interest: the walnut-shaped lump of tissue at the back of the brain called the cerebellum. In particular, the right half's 'crus cerebellum' (RCrusI) appears to be a little different in those diagnosed with ASD.[6] US researchers have chemically stimulated the cerebellums of genetically engineered mice, changing the firing rate of certain nerves in this area and 'rescuing' the mice's social behaviours. Granted, that's not quite the same thing as curing autism in people, but it does suggest that slow firing by specific nerves in the RCrusI might play some part in the difficulty many autistic people have in responding to social stimuli.

Engineered mice also played a key role in a discovery made by another team of US researchers, one that involved an area of the midbrain called the ventral tegmentum. Specifically, they found that having too many copies of a gene called UBE3A increased the likelihood of having seizures that preceded the loss of sociability.[7] By making special receptors for these nerves, the researchers found they could toggle the mice's social behaviours, effectively switching the autism-like characteristics on and off.

None of this is to say we're even close to some form of sociability treatments. Instead, studies such as this show the potential role certain genes play in the development of autism-like conditions. A variety of genes have been implicated in having a role; in fact, the current consensus is that as many as nine out of ten cases of autism can trace their cause back to an inherited factor.[8]

There's evidence that some of the genes responsible for ASD – at least in a significant proportion of cases – are inherited, while others are the result of what we call *de novo mutations*: changes in the genetic code that happen between generations.[9] Those new mutations tend to be responsible for impeded motor and cognitive skills, while difficulties with social and communication skills are caused by inherited genes.

Even if genes are ultimately to blame, it still leaves plenty of room for environmental factors, such as high levels of pollutants in the surrounding air or even particular gut microflora, both of which have been shown to potentially nudge up the risks.[10] Chemical speedbumps made of carbon and hydrogen stick onto segments of genetic code in what's referred to as epigenetics, interfering with the code's expression. These modifications can be added as a response to stress or other changes in the surroundings, effectively changing how whole networks of genes can function. Evidence is mounting that many cases of autism are not caused by genetic mutation, but are the result of the same kinds of genes being turned off by an epigenetic edit.[11]

Take-home message? Autism isn't a simple disease. Studies of genes and their expression are also revealing strong overlaps with the fundamental biochemistry of schizophrenia.[12] In other words, the two conditions merge like a fuzzy-edged Venn diagram, making it hard to say where one disease ends and the next begins.

We're only just beginning to learn how the expression and modification of genes depends on events in our surroundings. Just how those genes result in the pruning of synapses into diverse configurations is something we are barely able to grasp, let alone predict with any accuracy. The hints are there, though, and for the first time we're able to see exactly what the autism landscape looks like, even if the details remain hazy.

The fact that autism is diagnosed in up to five times as many boys as girls is still mysterious. Social expectations no doubt play something of a role, with examiners perhaps interpreting some borderline cases differently, based on their assumptions concerning how boys and girls ought to behave.[13] But there seems to be more going on than a simple gender bias.

One possibility lies in differences in the relative thicknesses of the cerebral cortex in boys and girls. Knowing there was a relationship between masculine brains and thin cortices, German researchers recently identified a correlation between the depth of the brain's wrinkled outer layer and ASD.[14] In other words, having a relatively thicker cortex pushed up your chances of developing traits defined as autistic. Women already form a slightly thicker cerebral cortex, so it's not a huge leap to imagine they possess some factor that helps reduce the impact of those changes.

In the near future, parents like my wife and myself might have access to genetic and lifestyle screens that not only tell us how likely it is that any offspring will have ASD, but also place them on a position on the spectrum. Perhaps we will even devise early prenatal testing: blood taken from the mother a couple of months after conception could be analysed for stray embryonic DNA, and its biomarkers compared with other risk factors to generate a prediction of how likely it is that the gestating child will be autistic. We can already detect foetal DNA that's slipped into the mother's bloodstream. Being able to test this material for any genetic condition we can think of isn't far away – an ever-growing pool of studies is leading to improved strategies that make it all but inevitable.[15] There will come a time when having a full genomic portrait of your unborn child will be as commonplace as the black-and-white sonar scans.

What should a parent do with this information? What would I have done? If magically granted foresight of our lives today – of having hit the jackpot with my lad – I'd have done nothing, of course. But if blind to the spectrum of possibility, and fearful of the possibility of a more challenging life as a parent, might I have made another choice as I held that genetic scan in my hands? I honestly can't say.

What is hypothetical for ASD isn't for numerous other conditions, from Down syndrome to spina bifida, which can be easily and confidentially diagnosed in prenatal screens. For decades, genetic counselling has been available to prospective parents who are aware of their family's connection with a disorder, especially where the genetics involved are relatively simple and bold in their expression. This kind of foresight is intended to inform reproductive decisions, preparing parents to make hard choices on how – or, on occasion, even whether – to proceed with the conceiving, birthing and raising of a child.

Prenatal screening can be seen as a way of improving public health, giving parents time to adjust and access the resources they'll need to raise a child with additional needs. Advances in genetic sequencing technology will squeeze more information from safer tests, at lower costs – an idea welcomed by many parents who are concerned about passing on debilitating conditions. But on the flipside, there's a concern that this kind of foresight – whether intended or not – risks bringing us to the edge of a slippery slope, one that historically we've often descended. It's a slope we will almost certainly slide down again in the not-too-distant future.

# GOOD

Entering our chubby little bundles of gurgling puke and pride into baby photo competitions is a rite of passage for many expanding families across the modern world. We all feel our precious angel deserves the best start in life with a first-place certificate sponsored by a name-brand infant formula company, after all.

In the United States Midwest about a century ago, it wasn't just junior who went into these kinds of pageants. And it wasn't so much about beauty – or not exclusively, at least. Ma, Pa and the whole brood could be judged in something called a 'Fitter Family' contest – the human equivalent of 'best in show' for pets and livestock at county fairs.

Emerging from Iowa State Fair's 'Better Baby' contest, the inaugural Fitter Family for Future Firesides event was conducted at the Kansas State Fair in 1920, under the careful scrutiny of one Mary T. Watts and University of Kansas professor of child care Dr Florence Brown Sherbon. In Watts's words: '[W]hile the stock judges are testing the Holsteins, Jerseys, and whitefaces in the stock pavilion, we are judging the Joneses, Smiths and the Johns.'

Entry wasn't quite as straightforward as submitting last Christmas's clan portrait with grandma craftily cut out. There was a medical exam, for one thing, which involved a check for syphilis (read: a morality scan). Then there was a psychological assessment, and the provision of a kind of family pedigree. Each family was scored with an alphabetical qualifier, with anything higher than a B+ earning you a bronze medallion inscribed with the words 'Yea, I have a goodly heritage'. Just the thing to lord it over the Joneses, the Smiths and the Johns at the next community picnic.

The leap from judging infants to evaluating the worth of a whole family didn't pop out of nowhere, mind you. We can thank a man by the name of Charles Davenport, who, after hearing about Watts's Better Baby contest, reached out with a postcard bearing a single sentence: 'You should give 50 percent to heredity before you begin to score a baby.' Not one to rush things, about a year later Davenport followed up with an equally cryptic comment: 'A prize winner at two may be an epileptic at ten.'

Though succinct, Davenport's postcards had their intended impact. An inspired Watts set out on a path to transform Better Babies into a family affair. After all, just how *can* you tell if a baby is really all that impressive until you've seen what it might grow into, especially since there was no shortage of prize-winning tots who were taken home by alcoholic fathers or neurotic mothers? After six years of planning the protocols, which were based on expert opinions, Fitter Families for Future Firesides was ready to be presented to the general public. Despite predictions that upstanding parents would be reluctant to take the necessary tests (not to mention fearful of what a Wassermann syphilis test might expose), twenty families entered that first year.

The contest quickly spread, with exhibits popping up over the following years in other towns and states. The tone wasn't always celebratory as much as cautionary: a display in a 1926 Philadelphia fair stated, inside a border of flashing lights, that 'some Americans are born to be a burden on the rest'. When we learn that the driving force behind the contest was the American Eugenics Society, this message probably isn't all that surprising.

*Eugenics* has become a word heavily stained by atrocity, being associated with totalitarian practices of forced sterilisations, if not outright genocide. It's an awkward word, and one we'd sooner leave in the past. But the rise and fall of its goals and founding principles is worth understanding, especially as its appeal has never fully faded, and more than likely never will.

As an ideology, eugenics has a history as old as civilisation itself. Ever since we discovered it was possible to grow more productive stock by only allowing the best to breed – whether wheat, cows or sows – we've understood that we can in some way shape the health and fitness of future generations of plants and animals. On a human level, our tribal inclinations have also encouraged a sense of 'us and them', and a belief in our strength and their weakness. For millennia we've wanted more of the good kind of crops, livestock and people, and fewer of the bad. Selective breeding, whether explicitly by practice or implicitly by custom and law, has been the answer.

But it took a perfect storm of advances in science and mathematics to turn hazy principles of selective breeding into a movement dedicated to the shaping of humanity at large. Following Charles Darwin's famous explanation of how species evolve under selective pressure, there came various academic musings on how his theory might apply to humans. One speculator was Darwin's own cousin – the famous biologist and statistician Francis Galton – who used his studies to conclude that since physical, intellectual and indeed moral traits seem to run in families, giving the upper class a leg-up over the hoi polloi, then those characteristics were to a significant degree biologically inherited.

Galton coined the term *eugenics* in 1883, drawing on Greek words that more or less translate into 'well bred'. This promoted the view that it's possible to breed desired characteristics into populations of humans, just as we have done for centuries with dogs, cattle and sheep. In his autobiography he described the burgeoning field of eugenics as 'the study of agencies under social control that may improve or impair the racial qualities of future generations, either physically or mentally'.[16]

Eugenics was really more than a simple study of those agencies: in a short time it was a political force, mixing a generous dose of assumptions with developing theories on natural selection and genetic inheritance, and insisting we use them to breed a better human species. This pressure tended to take two forms. Galton argued in favour of a positive form of eugenics, where rich and high-born bloodlines were encouraged to have more children and outbreed the 'more fertile' lower classes. The less benevolent flavour of eugenics was to restrict the ability of select members of the community to procreate, either through applying disincentives or, in extreme cases, through sterilisation. Or worse.

Galton's positive eugenics won little favour in his home nation, at least compared with the reception of his ideas across the Atlantic. America proved more fertile ground for his proposals. Two years before Galton originated the term *eugenics*, an engineer renowned for his role in inventing the modern telephone, Alexander Graham Bell, came to the conclusion that deafness ran in families. In a lecture, Bell suggested that it wasn't in society's best interests to allow those who were hearing-impaired to marry.[17] Or, more bluntly, to have kids.

Fuel for the eugenics fire came in the early 1900s, with the discovery of decades-old documents written by a German friar named Gregor Mendel. His historic experiments affirmed the results of contemporary studies into units of inheritance, providing the emerging field of genetics with solid mathematical models and turning it into a bona fide science.

As happens with most new areas of science, the shiny new lens of genetics was quickly used to examine all human mysteries. Novel assumptions about the gene's ability to cement the fate of an individual's biology were enthusiastically combined with Galton's eugenics, with the view of breeding in more 'good' genes and cutting out all the 'bad' ones. A renowned American pioneer of this union between population genetics and eugenics was an agriculturalist, zoologist and chicken breeder from Connecticut named Charles Benedict Davenport – the same man who would send those inspiring postcards to Mary Watts.

In 1910 Davenport founded the Eugenics Record Office in Cold Spring Harbor Laboratory, New York, which was tasked with gathering and recording the social and biological details of the US population. Its mission was simple: collect information on the ancestry of American citizens and use it to build a case for advancing eugenics. In the office's own words, the aim was 'to improve the natural, physical, mental, and temperamental qualities of the human family'.

Over the course of nearly three decades, the office collected details on hundreds of thousands of families, using interviews and records from hospitals and prisons to determine levels of fitness. Everything from 'feeblemindedness' to 'criminality' was assumed to be dictated in its entirety by the laws of population genetics, meaning any traits that weren't desired were mapped in the same way Mendel had mapped the colours of flowers and the wrinkled skins of peas.

The imperative of reproductive control was promoted by the proponents of eugenics as a way to ease the social impact of mental and physical disability, conditions that were unsympathetically described in terms not of individual suffering but of the cost of social resources. This was a far easier sell in a time when the voting public were increasingly concerned with state economics.

The community bought it: by preventing 'undesirables' from having children, citizens would save millions in future welfare and health care. On 7 March 1907 Indiana became the first state to enact a bill making the forced sterilisation of select individuals in state custody mandatory.[18] All up, thirty-three states would eventually introduce some form of sterilisation program, most targeting individuals who had anything from blindness to epilepsy. The legislation was not without controversy, with many bills scraping over the line following heated debate; some were defeated and sent back to the drawing board. Indiana's program was ruled unconstitutional in 1921, though the decision was flipped back again six years later. It was only put to rest for good in 1974.

For all the reported science and appeals to costs, a significant number of US citizens still found the idea of forced sterilisation repugnant. Even in states with laws permitting its use, the number of cases of actual sterilisations remained low – at first.

A 1927 Supreme Court trial changed everything.[19] To test the strength of a newly passed sterilisation law, the superintendent of the Virginia State Colony for Epileptics and Feebleminded petitioned his hospital's board over the involuntary sterilisation of an eighteen-year-old woman named Carrie Buck, who – allegedly like her own mother – was reported to have a mental age equivalent to that of an eight-year-old. Buck's guardian appealed the petition and drove the case up the judicial chain. After a succession of trials, the eventual Supreme Court decision went eight to one in the board's favour: Carrie was found to be feebleminded and promiscuous, and was therefore to have an operation to prevent her from having any more children.

*Buck v. Bell* set a precedent that permitted those in authority – not just legally, but morally – to deny patients the right to bear children. Consequently, a floodgate opened, with more than 60,000 individuals sterilised across the United States in the decades that followed. Most were women deemed to be intellectually unfit to be mothers, though the boundaries around what made for an intellectually fit parent were far from clearly defined. Being poor, black, indigenous or an immigrant often meant someone was more likely to be found to be unfit.

Where white American families faced near insurmountable obstacles in gaining access to contraception or voluntary sterilisation, minorities in the United States, especially people of African-American descent, were aggressively targeted by family planning initiatives. In many cases active consent for sterilisation was hazy or coerced; on occasion it was simply absent. In southern states, the overhanging threat of losing welfare or access to medical care meant African-American women routinely underwent tubal ligation, if not full hysterectomies. In 1961 the civil-rights leader Fannie Lou Hamer was subjected to what she called a 'Mississippi appendectomy' when, without her consent, she was given a hysterectomy while having minor surgery to remove a tumour. 'In the North Sunflower County Hospital,' Hamer explained to an audience in Washington DC, 'I would say about six out of the ten Negro women that go to the hospital are sterilised with the tubes tied.'[20]

That's not to say shocking abuses of trust didn't occur in the north, too. It wasn't unknown for medical students to train by performing unnecessary hysterectomies on African-American, and often poor, female patients. The Chicago Committee to End

Sterilization Abuse claim to have been told by an acting director of a municipal hospital in New York in 1975:

> In most major teaching hospitals in New York City, it is the unwritten policy to do elective hysterectomies on poor, Black, and Puerto Rican women with minimal indications, to train residents ... at least 10% of gynecological surgery in New York is done on this basis. And 99% of this is done on Blacks and Puerto Rican women.[21]

The problem was not confined to the United States. Many nations, both in Eastern and Western cultures, have had a history of removing the reproductive rights of their poor, kidnapping and relocating their indigenous, and sterilising their disabled in the name of saving costs and improving the prospective health and welfare of future populations. Australia has its shame of a stolen generation of Indigenous children. Japan has its *Kokumin Yūsei Hō* – a national eugenics law passed in 1940 that prevented the birth of disabled children. Sweden had a sterilisation program that began in 1935 and lasted four decades, sponsoring nearly 63,000 procedures for children considered mentally disabled.

The sun was slow to set on the field of eugenics. Over the latter half of the twentieth century, it gradually decayed as a legitimate area for study. Explicitly, at least. Almost from the start, scientists began to comprehend the sheer complexity of human genetics: where Mendel's pea plant experiments seemed so simple at first, nature versus nurture has turned out to be a tangled feedback loop rather than a clear dichotomy.

Less charitably, many sterilisation programs were abandoned as less invasive means of birth control became commonplace – a contraceptive pill was cheaper and less risky than surgery, after all. Not to mention easier to swallow than the horror of involuntary dissection.

Ultimately, it took the most dramatic of events to cast Fitter Families contests and forced sterilisation in a light that even the most committed eugenics stalwart would consider unthinkable. In 1933 the Nazi government in Germany passed the Law for the Prevention of Progeny with Hereditary Diseases, the top of a slippery slope which took the principles and practices of eugenics to ghastly places. Roughly 400,000 German citizens were sterilised for being epileptic, mentally unwell or physically disabled. Darker events were to come. The Holocaust and its goal of eradicating undesirables – not just in the form of the physically weak and disabled, but entire vilified populations – saw millions die. The bleak conclusion of Nazi eugenics forced the heads of many of the world's nations to reflect on their own health and medical practices. Shortly after the end of the war, the Nuremberg trials brought to the world's attention how far humans could go to study human weakness and frailty in the name of science, shocking us into enacting new codes and laws that governed how we should conduct medical research ethically, laws that remain in force today.

Thanks to a shift in ethics that aimed to preserve civil rights across the social spectrum, eugenics is now – for most purposes, at least – a dirty word. The societies and lobby groups that once proudly boasted the word in their names have either adopted new titles or faded into history. Universities no longer offer studies in the field of selective breeding among humans. Events where parents or other authorities petition for the sterilisation of a minor are controversial enough to be considered newsworthy.

But even if the word is contaminated, its founding principles have been harder to leave behind. The fear of the poor, the foreign and the intellectually challenged multiplying and taking an unfair cut of social resources is never far from many voters' minds. We all like to think we're immune, but most of us have had a chuckle over misfortunate fools whose accidental death at their own hand earns them a nomination in the 'Darwin Awards'. Mike Judge's 2006 film *Idiocracy* satirises the idea that lowly educated and inherently unintelligent people breed more than smart (read: middle-class and upper-class) people, resulting in a future that Galton would probably have put money on. Both might focus on vague pejoratives rather than actual racial identities, cultural classes or disabilities – cardboard cut-out representatives of a family lineage of humans too stupid to be allowed to persist – but the underlying eugenic sentiment remains alive and well.

Eugenics in the twentieth century showed us what we're capable of in the name of making humans 'better'. The limits it was taken to shocked us into reflecting on what it means to be biologically fit and healthy, and who gets to make those decisions. Thankfully, it will be a while before we see its likes again, if we ever do.

Arguably the most damaging legacy left by the eugenics movements has been the public's loss of trust in the relationship between science and authorities. Even the most robust of studies on the links between genetics, heritage and cognition will continue to carry the baggage of half a century of eugenics propaganda. It's hard to discuss the impact genes have on intelligence without invoking memories of sterilising the mentally feeble or vilifying 'sub-human' races. Antenatal screening hovers on the boundary with eugenics; some think it sits well inside. The conversation is an emotional one, but that shouldn't make it off-limits.

Of course we must stay cautious, given our track record. We also need to be careful not to throw out a healthy baby with the Aryan bathwater. Revolutionary gene-editing technology has already shown that it's possible to erase inheritable heart conditions from a human embryo, under strict laboratory conditions if not yet in an applied setting.[22] Most other demonstrations of the technology's potential in curing disease has been limited to mice models, but could soon see an array of human treatments for everything from HIV to cancer.

It's all thanks to a new kind of molecular tool called CRISPR. It stands for Clustered Regularly Interspaced Short Palindromic Repeats – sequences of genetic code inside bacteria, which make up a library of viral mugshots. Coupling this list with one of a number of enzymes called a Cas (short for 'CRISPR-associated protein') makes for a serious search-and-destroy missile, one that targets a virus's genetic sequence with pinpoint precision. Once inside, the Cas can gently snip the sequence or woodchip the code into oblivion, depending on which kind of enzyme you've used.

This handy way to target a specific section of DNA and use an enzyme to break it apart makes for an incredibly useful tool in genetic editing. And it's not the only tool in the kit – other processes are emerging that can gently pull out single bases, like the world's tiniest dentist popping out a nucleic acid tooth. The ability to tweak our genetic code with the aim of 'fixing' disorders is quickly becoming less a practical issue than an ethical one.

Editing genes to remove what we see as faults all boils down to what exactly is considered faulty. Fundamental to this is the question of what we think of as 'good' versus what we see as 'broken'. Logically, *good* is not a quality that can be found in isolation in nature. Philosophers call the misunderstanding that there's some sort of objectively desired format of objects in the universe the *naturalistic fallacy*, but we can also think of it another way: Mother Nature does not give a toot about things like life and death, sin and virtue, pain and suffering, happiness and misery, and so on. She doesn't play favourites. 'Good' is not a property of physics and chemistry; it's what we humans value within a specific context.

That said, some desires are fairly universal among humans. As self-aware biological entities, most of us meat-bags are averse to pain, mental suffering and the deprival of relative amounts of freedom. We seek comfort, pleasure and – as my old school principal used to say – a hassle-free day. As social entities, most of us hate to witness suffering, or to lose our ability to connect with loved ones. We usually like to hear the laughter of our children, and know our parents are comfortable in their old age. No matter what language you speak or how much pigment your skin has, those are likely to be truths.

Keeping the naturalistic fallacy in mind, 'bad' mightn't necessarily be completely arbitrary either. It's hard to imagine that a mutation in the gene for a protein channel resulting in cardiac arrhythmia or cystic fibrosis, and ultimately risking an early death, would be a desirable characteristic for anyone. Removing the repeats that cause Huntington's disease could give a fertilised embryo a longer, more active life. Gene-editing through technologies such as CRISPR has enormous potential to bring an end to debilitating conditions that cause enormous levels of pain, mental anguish and early mortality – and that, let's face it, would generally be regarded as a good thing.

Reducing levels of melanin in the iris, on the other hand, is more a matter of taste or cultural bias, making it much harder to unanimously categorise as either good or bad. Different cultures have preferences for ratios and orders of genders among their children – that's another characteristic that isn't likely to be universally counted as good or bad, and yet could be modified through genetic screening, editing and selection.

There is no clean line between disease and desired. Nor should we pretend there is. Decisions to edit genes – decisions that could prevent great suffering – shouldn't be led by the impression that science can highlight what to cut and what to keep.

To make matters complicated, genetics is rarely straightforward even in principle. Some detrimental mutations might call for a simple letter-swap. Other conditions could call for changes that risk serious complications downstream, or at the very least might tailor a personality in other unforeseen ways.

What if, by excising genes that make an individual prone to depression or anxiety, we also deprive them of greater levels of sensitivity or creativity? Maybe if we'd tailored our son's DNA to have one copy fewer of UBE3A, a slightly thinner cortex and higher firing rates in his RCrusl, he might find it easier to understand my jokes and not find thunderstorms so terrifying. But what if the cost was his passion, his rich imagination and his uncanny ability to find meaning in obscure patterns?

Darwin's model of evolution revolutionised how we understand diversity. At its centre sits this ideal of *fitness* – a trait that gives individual organisms the edge over their competition. We tend to think of these traits in direct terms – of strength, awareness, endurance and virility. But humans didn't take over the world with sharp teeth and strength alone. Our intellect, though impressive, also had little to do with it. Humans thrived because our social brains, working in combination, could create a superorganism of tribes.

Just ask Shanidar 1. Today he's a jumble of fossilised bones pulled from a cave in Iraq, but when he walked Earth's surface some 40,000 years ago, as a middle-aged Neanderthal man, he had a few problems. Shanidar 1's body was distorted, more than likely the result of paralysis resulting from a blow to the skull he received as a child. His body is remarkable as it provides clear evidence of social care being devoted to an individual who, we might expect, would have been left behind for the buzzards to feast on.

Perhaps he would have been abandoned in a time of famine. Maybe Shanidar 1 had other skills that made him an anomaly worth holding onto. We can only speculate. But, taken at face value, his bones are a reminder that the family tree of human animals evolved to compensate for physiological losses of one kind or another in order to maintain the all-important group bonds. It doesn't matter if you can't make babies or hunt mammoths if you can provide child care, wisdom or social support. Fitness isn't simply about an individual's robust organs. Darwin's idea of natural selection doesn't fall apart when we look after our ill, our disabled and our elderly. It's strengthened.

Eugenics is a ruined field of study, likely to forever remain buried in our past. Yet its ideals are hard to run from as advances in biology and medicine result in novel technologies promising an end to suffering in the same way antisepsis, antibiotics and vaccines have done. We face the challenges of deciding which biological traits are good and which are bad, who should decide the difference, and who can access the resources necessary to make a difference. To meet these challenges, we first need to know what we mean by *disease*, and distinguish it from its moral underpinnings. We must clearly articulate what we believe to be good and bad, free of the assumptions that have blindly led us to commit such atrocities in the past.

# FERTILE

How to make a human, according to alchemist Theophrastus von Hohenheim (a.k.a. Paracelsus), circa 1537:[23]

Step 1 Putrefy a generous amount of sperm in a gourd of some sort for forty days.
Step 2 Once it's good and stinky, place the gourd into a horse's womb until the mass of sperm is clearly squirming.
Step 3 Check to see that it looks a bit like a person, only transparent.
Step 4 Feed your transparent man-baby some human blood for the next forty weeks.
Step 5 Pop the gourd back into the horse's womb until the human is done.

Makes one baby.

Back when I assisted in the production of a baby one Valentine's Day a few years ago after a lovely teppanyaki dinner, I used a more traditional method than von Hohenheim's. This was largely because a) we had no gourds, b) it was a heck of a lot more fun, and c) we had a hard time convincing the RSPCA that the horse was a willing participant. My partner and I expected the fertilisation part would be much harder than it was, but after a single slightly disappointing menstrual cycle, the magical striped stick said we could begin the nervous nine-month wait for our brand-new human.

We were fortunate. The exact statistics depend on things like age and health, but fewer than half of couples trying to conceive are successful within the first three months. Around nine out of ten make a baby within the first year. For many wannabe parents, however, the dream of having a family fades into heartbreak as the months roll by. Roughly 9 per cent of couples qualify as infertile.[24]

Since the late 1970s, the medical technology called in-vitro fertilisation has provided them with one last shot at parenthood. A staggering 350,000 children worldwide are now conceived and delivered using this process each year, meaning a reasonable percentage of our global population started life with some gentle guidance inside laboratory glassware.[Δ]

We can expect this figure to climb even higher – not only because of a growing world population, or because the technology continues to improve, but because infertility is on the rise. We're simply not pumping out quality sperm like we used to. One recent study carried out by the Hebrew University of Jerusalem on 43,000 subjects found that the average sperm count across Western nations is dropping by about 1.4 per cent each year. Between 1973 and 2011, concentration of our little swimmers fell by 52 per cent. And we don't have the foggiest idea why. Everything from tight pants to pollution to mobile phones has been blamed, but there's little concrete evidence.[25]

There is something of an irony in this. For most of history infertility has primarily been a woman's disease, with the term *barren* being used to invoke the impression of a desolate field where seeds won't grow. A man could be impotent, of course, but so long as he could still ejaculate, he wouldn't be considered sterile. In the 1850s an American gynaecologist named James Marion Sims experimented with one of the first examples of assisted reproduction in humans. His failure to artificially inseminate his patients probably came down to two things: he didn't take into account fertility cycles in women, and – more likely – he never stopped to consider that the patients weren't falling pregnant because the father was shooting blanks.

---

Δ  The International Committee for Monitoring Assisted Reproductive Technologies estimated that there were around 237,809 ART babies born in 2004, a 2.3 per cent increase on the previous year. Allowing for reasonable increases in technology, more than 350,000 is a fair estimate.

It was only a few decades later that we have the first successful example of a woman falling pregnant following artificial insemination with sperm from a donor. Not that she was told ... A Philadelphia physician named William Pancoast realised that his 31-year-old patient who was desperate to conceive was failing not through any fault of her own, but because her husband's semen was 'absolutely void of spermatozoa'. To remedy this, he took a semen sample taken from one of his medical students and inseminated his patient while she was knocked out with chloroform during what she presumed was a routine gynaecological examination. After she delivered a healthy boy nine months later, Pancoast revealed the truth to her husband. Together, they agreed it was probably best to keep the procedure a secret from the patient.

What has been a medical condition could soon, with further slipping of boundaries, become a new norm. No doubt there are plenty of people who would cheer the flatlining or even decline of our planet's human numbers in the name of conserving resources or preserving the environment. But putting aside the sustainability debate, the question of who should have access to fertility treatment is an important one. With more men becoming infertile, should only those with sufficient wealth have access to a solution?

Francis Galton would, of course, say yes, being both mistaken in his belief that the poorest generally outbreed the richest, and biased in his view that there is an inherent good in the upper classes spreading their genes. Now, in the twenty-first century, it's surely time we gave up thinking that having money reflects deeper virtues.

In Australia, at the time of writing, it will cost you around $2500 to have a trained professional insert sperm into your uterus (referred to as intrauterine insemination), giving them a good running start. Throw in a generous rebate from socialised medical welfare and you'll probably pay just under $2000.[26] If you need a little more help, a single cycle of IVF costs just over $9000. Again, with Medicare, you'd only be out of pocket $3000 to $4000. But perhaps a single cycle won't be enough – and then there's the hospital fees, and maybe the need for an intracytoplasmic sperm injection, where a single 'chosen one' sperm is helpfully nudged into an ovum using an extremely fine syringe. Head over to the United States and it's a slightly more depressing story. While the treatment costs range from similar to virtually double, your out-of-pocket expenses will depend on your insurance.

The end result, however, is much the same throughout most of the world: if you can't afford to buy a second-hand car or go on a nice holiday, you probably won't have the funds to overcome any biological hurdles you face in making a baby. And statistics show around a 16 per cent decrease in access for the poorest 20 per cent of the Australian population.

So many words have been shared through history on the importance of keeping the poor childless that it's virtually impossible to have a dispassionate conversation about assisting reproduction among those in lower socioeconomic brackets. The arguments primarily focus on the costs of raising a kid. A report from the National Centre for Social and Economic Modelling in Australia argued that the cost of getting two kids to their twenty-first birthday is around $812,000.[27] A similar report by the US Department of Agriculture found that for a middle-income family to raise a child born in America in 2015, they'd spend US$233,610 between birth and age seventeen, give or take, which puts them in the same ballpark.[28] So if poor people can't afford IVF, there's no chance they'll have enough money to feed, house, educate, treat and entertain any kids, right?

Ignoring the multitude of problems with this kind of thinking, such as the variation in costs that make those numbers *averages* rather than *limits* for the raising of an educated, happy, healthy kid, the bottom line remains starkly clear: overcoming infertility is something we as a society consider a privilege. Granted, this isn't unique to the field of assisted reproductive technologies: all kinds of treatments are far more easily available to those who have money. So fertility isn't special, but it is a bold illustration of the perfect storm we often sail into when melding the moral and the medical, and the debate over whose biological differences deserve public help versus those whose difficulties are a consequence of choice.

The current record holder as the world's oldest mother is thought to be Daljinder Kaur from Amritsar, in north-western India. She is alleged to be about seventy-two – a controversial age to have been granted IVF assistance in anybody's books.[29] A decade younger, at sixty-two, is the unnamed Australian woman who hit national headlines in 2016 when she went overseas with her husband to receive a donor embryo, before returning home to carry the child to term.

While most media reports stuck to the facts, others were far more opinionated. A story in the conservative tabloid *Herald Sun* quoted the perspective of a respected pioneer in Australian IVF treatment, Monash University professor Gab Kovacs, who stated:

> I think getting people of that age pregnant is irresponsible. That child will need looking after for 20 years, and there's a possibility she won't be able to do that. Our bodies weren't designed to have children in our 60s. I don't think any responsible IVF unit in Australia would treat someone of that age, and it's not a standard of medicine I would condone.[30]

I'd never go toe-to-toe with Kovacs to argue the science behind human fertility and the progress of cutting-edge treatments. He'd certainly know his stuff. His words not only express popular beliefs about fertility, they are the view of a robustly informed and highly influential medical authority. In that one statement we can see what makes a disease a disease in the mind of a medical expert, and why people like the anonymous 63-year-old mother aren't felt to deserve medical assistance.

Most concerns take into account the welfare of a child under the care of someone whose health is likely to fail in coming years. Not to diminish the concern itself, which isn't invalid, but it's worth noting that we rarely discuss health issues of elderly fathers, whom we tend to congratulate for being virile rather than irresponsible. In other words, there is a moral order in who we think is responsible for a child, with the mother at the top and the rest of the community lower down.

There is also the question of gambling on continued improvements in quality of life. It's not unreasonable to think that in the next few decades, a good many people in their eighties will have a quality of life similar to those currently in their seventies. Our impression of a frail octogenarian shuffling around after their spry teenager might need a rethink, as the stereotype gives way to generations of elderly people who have more freedom than ever.

Another general concern addresses what we perceive as the normality of human biology – something summed up in Kovacs's statement that women in their sixties 'weren't designed to have children'. Assuming he isn't a creationist talking about divine blueprints, his statement concerns the medical distinction between menopause and infertility. One is natural and to be expected; the other is a disease. Crossing that line makes risks less acceptable and benefits less desirable.

Risk management is fundamental to being a medical professional, and intrinsic to the ideal of 'doing no harm'. It sits up against the inescapable reality that we must balance costs with chances of success. On the spectrum from policy to patient, there is a tightrope of health that's only as thick as the public's purse will allow it to be. And treating disease always comes first.

In 2006 the Australian federal government attempted to reduce welfare spending on IVF, particularly for older women. The argument from the conservative government was that as IVF was an elective, non-essential procedure, the public shouldn't have to contribute the level of spending it was currently. The plan was to limit subsidies for anybody over the age of forty-two to just three IVF cycles. IVF Australia pushed back and the move was defeated.[31]

A decade later, Fertility Society of Australia president Michael Chapman returned to the question of whether older women should have their IVF treatments subsidised by Medicare, citing the imbalance of the mathematics involved. At age thirty-six, your chance of conceiving naturally is half of what it was at age twenty. By forty-one, it's a mere 4 per cent. If you add up the total cost to help women over the age of forty-five conceive and then divide by the number of babies born, you get a number close to around $200,000. For women aged thirty, that figure is closer to $28,000. 'Is that value for money?' Chapman asked.[32]

When we combine the fact that couples are delaying having children until later in life with the aforementioned decline in male fertility, the question becomes even more pressing. The average age of women undergoing some form of assisted reproduction technology is currently hovering in the mid-thirties, and it's slowly inching its way up. A quarter of all women using IVF in Australia are already over forty.[33]

Menopause isn't considered a disease. Neither is the natural decline in fertility that comes with age. Delaying the starting or even continuation of a family in favour of a career or freedom is a choice, not a consequence of broken biology. Even if medical science overcomes the hurdles that come with having children late in life, concerns about risk, cost and benefit to the community won't go away. Chapman's question will always remain pertinent: *is that value for money?* We must be well prepared with a good answer.

The medical world has moved beyond the castigation of sexuality and gender identity as diseases in their own right, even if select pockets of society have been slower to come to the same conclusion. But if being gay or transgender isn't an illness, how do we medicalise what can effectively be infertility, given that we've historically defined fertility by the involvement of an erect penis plus testicles, a welcoming vagina connected to a functioning uterus, and some Barry White on the stereo? How do we justify using medical technology to assist those who wish to be parents yet are barred by their biology?

In some respects, the question is as outdated as it is philosophical. Women have used assisted reproductive technologies to raise families without a male partner since the start, procuring the necessary gametes from services like the Sperm Bank of California as far back as the early 1980s. Male couples have made use of surrogates for egg donation and gestation, looking beyond the traditional to define what it means to be a family.

There are also options for transgender folk who are about to embark on transition procedures that will see them lose reproductive tissues: they can currently make use of assisted reproductive technologies to preserve their sperm, ova or embryos. In the absence of discriminatory laws excluding individuals from making use of IVF services, and in light of recognition acts that give transgender men and women equal rights, the processes available to heterosexual couples are readily available to those of all genders and sexual flavours.

Beyond workarounds, piggybacking and loopholes, assisted reproductive technologies that explicitly help individuals outside of traditional family structures aren't exactly a high priority. Right now, transgender women can't experience pregnancy and deliver a child. Transgender men can't inseminate a female partner with sperm containing their own DNA. Gay couples don't have ready access to technologies that might combine their genes in a single embryo; tangentially, polyamorous partnerships consisting of more than two individuals have no way to mix their genes and share in the production of a new life.

There are potential technological solutions on the horizon. The first transgender woman scientifically recognised to breastfeed her newborn made headlines in early 2018. Her lactation followed an experimental treatment consisting of an anti-nausea drug, an array of hormones and physical breast stimulation.[34] The first child to gestate to term in a transplanted womb was born in Sweden in 2014, and a number of successes have followed.[35] A near-term lamb foetus had its last few weeks of gestation completed in an artificial womb developed by researchers at Children's Hospital of Philadelphia in 2017, suggesting that treatments for premature births might slowly bridge into systems that could raise an embryo outside a human body.[36] Could transgender women one day use these kinds of technology to give birth? What about cis-gender men?

Great leaps have been made in transferring the nucleus of a body cell into egg cells and encouraging them to develop into embryos. Early 2018 saw the world's first healthy delivery of monkeys cloned using the same somatic cell nucleus transfer that was used to clone the famous sheep Dolly.[37] That brings us one step closer to engineering human embryos by combining the chromosomes of same-sex parents, or potentially even sourcing genes from multiple parents.

The first child to contain DNA from three parents was born in 2016.[38] The mother had a condition called Leigh syndrome, which has been traced to genes in the mitochondria. An ovum from a third 'parent' provided non-maternal mitochondrial DNA. A year later a similar technique was used to treat infertility in a Ukrainian couple. The procedure is still in its infancy, and not without controversy over its risks and benefits.

All of this progress adds up to the potential for more opportunities, helping more individuals share in the ability to make children with those they love.

While the research behind these recent milestones will no doubt assist a great diversity of needs, fixing disease and disorder remains the primary focus. Nuclear transfer is to treat Leigh syndrome or infertility; womb transplants were intended for women with uterine abnormalities. We research medicine to fix the sick, and this intention informs the public's interpretation of its potential – which in turn defines the ethical boundaries of its application, and finally the laws framing its integration into medical practice.

Medicine follows the credence that 'if it ain't considered broke, don't fix it', leaving politicians and researchers and opinionated editorials to debate rights and values, while maverick bioengineers and fringe surgeons boldly – perhaps arrogantly – redraw the limits on how this technology can be applied. Even once these new medical boundaries are accepted, Chapman's bluntly honest and necessary question – 'Is that value for money?' – will continue to discriminate between those who can easily access revolutionary treatments and those who require assistance from public funds.

How we define disease matters when it comes to tinkering with the human body. If you're broken, we can repair you to a standard of health; we can take risks, seek support and afford you charity in order to bring you back to normal. If you're healthy, the change is your choice, so it's on your dollar. Disease sets the limits of change we find acceptable, beyond which we become something different. Something more than human, yet at the same time – paradoxically – something less than human as well.

# FIT

The Australian Institute of Sport isn't all that far from where I live. I often cycle through its quiet, leafy grounds as a short cut into the city's centre, passing the halls and grandstands and running tracks where Australia's sporting elite train their bodies and minds to peak performance under the guidance of expert coaches and scientists.

In front of the institute's visitor centre there's a magnificent metal sculpture of an athlete reaching up out of a wheelchair, arm outstretched towards a ball. I'm neither an art critic nor a sports fan, but to me the sculpture's message seems clear: *This is where people push themselves to their full potential.* I like that statue.

Perhaps there's another message being presented, though. An unintentional one, but just as poignant: *This is also where technology pushes a diversity of human bodies to their full potential.*

Long before he hit headlines for the murder of his girlfriend, Oscar Pistorius was the South African sprinter known as 'the Blade Runner'. He had been born without fibulas (the lower leg bones), leading to amputations while he was still an infant. Pistorius quickly mastered the prosthetics that replaced his lower legs, and earned notoriety for competing in single below-knee amputation sprints – classified as T44 events – in spite of having double below-knee amputations.

This wasn't the only controversy early in his career: beaten to the gold in the 2004 Summer Paralympics in Athens by Brazilian runner Alan Oliveira, Pistorius complained that his competitor's longer prosthetics gave him an unfair advantage. 'Not taking away from Alan's performance, he's a great athlete, but these guys are a lot taller and you can't compete [with the] stride length ... We aren't racing a fair race,' he said.[39]

The statement would become rather ironic. Pistorius's real ambition was to compete alongside athletes who still had the legs they were born with, a goal he realised in 2005 when he took sixth place in the 400-metre South African championships. But the question of whether his anatomy gave him a boost in speed and endurance was yet to be definitively answered.

Studies conducted on Pistorius and his J-shaped Flex-Foot Cheetah 'blade' prostheses suggested he spent anywhere between 17 per cent and 25 per cent less energy compared to athletes running on old-fashioned forms of human anatomy, did 30 per cent less mechanical work in lifting his body, took 21 per cent less time to swing his legs each step, and had a foot in contact with the ground 14 per cent more often.[40] What this added up to in terms of advantage was debatable, but the International Association of Athletics Federations was convinced, and banned Pistorius from competing against non-amputee runners.

An appeal saw the ban lifted, and in 2012 the South African Sports Confederation and Olympic Committee announced their nation's Olympic team would include the Blade Runner. Oscar Pistorius would compete in the 400-metre race, and the 4 x 400-metre relay, becoming the first amputee to compete in an Olympic event. He didn't come first, but still outran other competitors.

Exactly how the racing community would have responded had he won is open to speculation. No doubt it would have sparked a heated debate that would have shaken up how we define not just the nature of disability, but sport in general. Some years later, the German Paralympic long jumper Markus Rehm would be denied entry into the 2016 Rio de Janeiro Olympics based on his inability to demonstrate that his own blade prosthesis wouldn't give him an unfair edge on other jumpers. It's evident there is still confusion over where disability ends and technology begins.

Competitive sport pits body against body, and brain against brain. It's a match of biological strength and human psychology.

It's why there are separate categories for ability, clear rules and distinct levels of competition. Our enjoyment of sport comes from the tension that comes with meeting an opponent who presents a decent challenge.

In some sports, we accept the necessity of technology. Teams of engineers work hard to research and apply the materials and mechanisms that will win the day. Motor sports might celebrate the minds and bodies of drivers, but their teams of pit crews and car designers aren't far behind. Nobody would believe for a moment that the athlete in the seat wins the race all on their own. To a degree, we also accept that technology must play some kind of role in other sports. Try running without shoes, hitting a ball without a bat or racquet, or shooting an arrow into a bullseye with a medieval longbow. Where materials are involved in some way, there will be scientists striving to maximise performance.

It's not always clear where the line is between helping and augmenting – that is, when performance is more down to technology than biology. In 2012 the Australian swim team presented their new Olympic weapon. The Fastskin suit was developed by Speedo in the early 2000s. The fabric it was made of was inspired by shark skin, specially designed to repel water, and constructed to hold and compress the relative muscles of its wearer. The outfit clearly gave the swim team some kind of an advantage; Speedo had claimed the suits reduced drag by 4 per cent. Human plus Fastskin swimsuit made for a superior swimmer. So superior, it turned out, that the suits were banned by the regulatory Fédération Internationale de Natation for the London Olympics in 2012. The decision isn't a clear-cut one, though; fast suits have been permitted in other competitions.

The intersection of sport, disability and technology seems to mirror the junction of medicine and disease in some ways. Both see a boundary of human biology, and seek to apply knowledge and discovery to help people reach that line. But to go no further.

At the Australian Institute of Sport's visitor centre I met with a man by the name of Peter Downs, hoping he might help me better understand where society currently stands on the topic. For seventeen years he was the manager of the Australian Sports Commission's Disability Sport Unit, before moving on to manage Play by the Rules, a collaboration that provides information and resources to promote safe and inclusive behaviour in sport. If anybody could bring me up to speed on a topic I knew nothing about, it was Peter. It turned out that, as with health and medicine, the conversation around sport and technology is far from simple.

'Twenty, thirty years ago, you'd see people compete in wheelchairs that were little different to the ones they'd use in everyday life,' Peter told me. 'Now athletes are working with chairs designed to suit speed or distance. Look at wheelchair rugby, for example. The chairs are designed to handle the impacts, and to be safer and more manoeuvrable.'

'So it's a little like motor racing?' I suggested. 'You've got a whole team responsible for helping you win.'

'It is a little,' he acknowledged. 'It's an area constantly under improvement.'

And it's not just chairs. Athletes of a variety of abilities are using all manner of frames, blades, wheels and electronics to assist their physiology. Nobody would say the tech is doing all the work, of course. But without it, it's questionable whether even the fittest body could hold its own against athletes fitted with ultralight materials tailored to channel strength, endurance and power into an action. As with medicine in general, there is an ethical question concerning equity: who can afford access to a team of experts who might push their body just that little bit further?

When I was in my early twenties, a short piece of science fiction I wrote won a prize in a state competition. With hindsight I'd say the story was a little clichéd, but its premise has always amused me: a future Olympics where pharmaceutical companies compete to develop the best drugs, and push their sponsored athletes to their physical limits. I shared this story with Peter, who smiled politely and conceded that although it sounded very sci-fi, anything might be possible in the future.

'It's just one piece of the pie,' he said. 'Technology, regulations, rules. They all move with one another, helping sport to become more diverse and inclusive.'

It's not always so straightforward. Peter explained how he once delivered a talk to a Korean sporting delegation, the members of which seemed incredulous that we could simply 'change' the rules on what constituted a single sport. It made me think of so many other parts of society where the borders seem so immovable that even thinking of changing the rules to accommodate disability appears ludicrous.

This feedback loop between technology and sport's boundaries has important consequences that go beyond the realms of competition. Engineering that assists athletes often finds its way into improving prosthetic limbs and other supporting devices, making them lighter, stronger, more comfortable and easier to use.

Stepping back, the conversation about technology in sport echoes the debate on health and medicine, even beyond questions of inclusivity. In both we have an expectation of what the archetype of a human body is capable of doing, and we develop technology that aims to meet that standard.

If having prosthetic limbs made us faster, how would we feel about amputating otherwise healthy limbs? Drugs in sport already tests our acceptance of enhancing humans beyond the limits of being human. Could my sci-fi story ever become a reality?

There is a subculture among us who see these limits as a challenge. Not so much in the realm of competitive sports, but in pursuit of aesthetics, of beauty, of leisure – and … well, in pursuit of pushing limits just because they can. 'Body hacking' is where medicine meets art. In some ways, of course, we've been hacking our bodies forever. The 5300-year-old remains of a mummified hunter we've dubbed Ötzi the Iceman was found with numerous parallel lines and crosses tattooing his preserved skin, showing that we've been poking ink into our flesh for a while now.

We've come a long way since Ötzi's stripes, of course – I myself have a large blue squid adorning my arm and shoulder. I also have metal rings poking through my ears and a gaping hole stretched inside one of the lobes. Why would I do such things? The most adolescent response I can muster is simply because I like it. The anthropologist in me understands it has to do with connecting with a tribal aesthetic, one that says adorning my body in such ways reflects values of beauty and strength, while setting me apart from others in the community I don't identify with.

We've been rebuilding body parts for those afflicted by battle wounds or the scars of pox and syphilis for centuries. As the risk of infection and the promise of pain has diminished, those willing to roll the dice under the knife have used surgery not just to replace, but to improve. We've built breasts, reshaped noses, sucked out fat and chiselled chins, all to have an appearance that suits our tastes.

These changes pale in comparison, though, with individuals who 'hack' their bodies in vastly more extreme ways, inserting implants that ripple their skin with nodules or curves, burning and scarring words and patterns into their flesh, splitting their tongues into forks, inking the whites of their eyes and so forth. Why would *they* do such things?

Beyond ideals of tribal affiliation and notions of aesthetics, some self-confessed body hackers interfere with their biology in order to change how it fundamentally works, all in the name of curiosity and advancing what it means to be human. At the more innocuous end of the spectrum, this could involve simple microchip implants and radio frequency transmitters slipped under the skin that can send and receive information, much like a key card that you can't lose. Or how about embedding a subdermal magnetic device that hums when introduced to an electromagnetic field, giving you the ability to tangibly sense an invisible landscape of light which nobody else can detect? The limits of what's possible are set by only two things: what an individual is willing to risk, and what materials are conveniently available to experiment with. And both are constantly evolving limits.

There will always be somebody willing to put one foot over the line, to cut a little deeper, stretch a bit further, try new chemicals, take heavier doses. Hardware will continue to miniaturise, opening the way to subdermal displays that can be read through the skin. Power sources running on kinetic energy, or perhaps even our own blood's dissolved glucose, will help biohackers explore new fields, just as they'll also progress the development of pacemakers or artificial organs. Drugs will be discovered or synthesised as quickly as they're banned.

Already there are signs of biohackers delving further into genetics. In 2017 a former NASA biochemist, Josiah Zayner, created a media stir by injecting what he described as CRISPR-modified myostatin – a muscle growth protein – into his arm during a live-streamed event.[41] It was a stunt, but an effective one, drawing attention to the next step in non-medical applications of genetic engineering. A few months later, the CEO of a biotech start-up company called Ascendance Biomedical dropped his pants onstage at a biohacker conference, to inject into his own thigh what he referred to as a 'research compound' that could treat, if not cure, a number of infectious diseases.

What a short decade or two ago took entire university departments millions of dollars to achieve is rapidly becoming achievable for pennies, using internet-purchased equipment in a spare room. Of course, you also need a brain experienced enough to make use of it.

Biohacking skirts the edges of what is legal and – from some perspectives – what might even be sensible. Technically, what we do to our own bodies is our own business, so long as we don't sell and promote therapies as untested treatments. While officially they might be ignored, unofficially the authorities are paying close attention to what is a medical frontier. There will be mistakes, but what better way to learn than to watch while simultaneously washing one's hands of responsibility?

I must admit, I have a certain respect for mavericks who seek to change what it means to be human. We actively encourage the improvement of our bodies, so long as it serves to meet a standard of normality. There is a wary acceptance of biohackers nudging their biology into new frontiers; it's begrudgingly sanctioned by authorities thanks largely to the potential benefits it has to improve disease and disability. Early in 2018, at a biohacking conference, the US Defense Advanced Research Projects Agency's biotech office director, Justin Sanchez, responded to a question about the agency's relationship with those who engage in the practice of pushing healthy biology past its limits, stating, 'Meeting people who also think about the future is a critical part of considering what's ultimately out there.'

We're tentatively tiptoeing our way into a future where normal is no longer a universal constraint, where we can hack our bodies not just to fix them but also to be different. Be it for beauty, sport, identity or simply to satisfy our own curiosity. It's a future where medicine isn't just about fixing something broken, and where technology will continue to test what we mean by good versus bad, and normal versus abnormal. So we have to ask: just how do we define *disease* in this kind of future?

# VALUABLE

Facebook founder Mark Zuckerberg has a lot of money. In 2016 he wanted to use some of it to cure every human on the planet of disease. Together with his wife, Priscilla Chan, Zuckerberg promised to devote US$3 billion from the philanthropic Chan-Zuckerberg Initiative to research technology that will one day help heal biological suffering.

Curing everything that afflicts humanity is a big call, and moonshots like these make great headlines. Even Chan had to agree with that. 'This doesn't mean no one will get sick,' she admitted in her speech to a room of people in San Francisco. 'But it does mean that our children and their children should get sick a lot less.'[42]

It's a goal that's as laudable as it is impossible. Not just 'this is really too hard even with billions of dollars' impossible. Not even 'some things are impossible to ever know' impossible. It's impossible because, as we've seen, the tides of disease ebb and flow with the generations, changing under the ever-present pull of our moral gravity. We might build new islands of health, but those shifting waters won't ever stop.

Human biology is diverse, and with that diversity there will always be functions and behaviours that cause conflict and distress relative to our levels of comfort and happiness. New technologies and processes will give rise to pressures we never imagined; just think of how dyslexia was barely an issue before the invention of writing. Or how motion sickness has become prevalent in the age of high-speed transport. How will diverse neurological traits that are hidden today manifest as debilitating inconveniences in the future, as we increasingly automate, digitise and migrate into new environments, perhaps eventually onto new worlds?

Even with greater control over our genes, our biochemistry and our microflora, we will invent novel ideas about what it means to be human and what levels of suffering we should tolerate. For as long as we have biological diversity, moral expectations and a sense of free choice, there will be disease.

But let's not stop the Zuckerbergs from throwing in buckets of cash! The point isn't that we don't suffer, or that suffering as a result of our biology is something to be tolerated. Indeed, fuck cancer and HIV. Let's find ways to eradicate the scourge of malaria and see a final end to polio and measles. I hope to live to see ways to treat or even prevent Parkinson's disease, Alzheimer's and other forms of dementia. Let's find treatments that do away with anxiety and depression, regardless of whether they're pills or therapies or a mix of both. Disease might be a tenuous, even outdated concept, but pain and anguish and loss of function are hard truths that demand our empathy, our focused attention and our resources.

Understanding what makes a disease a disease isn't mere pedantry. It's an acknowledgement that, when we think about words like *sick* and *better*, we have a responsibility to consider exactly what these words mean, and who gets to decide.

So just who does get to determine what makes a disease a disease?

To give a trite answer, they do.

To give an even more unoriginal answer, you do.

It's about the language we all use in our medical decisions, our voting selections and our media conversations. Disease is defined by a mass of tiny actions we all make every day, actions that tell politicians, journalists, researchers and physicians how to act when it comes to health and medicine. It's how we talk about the limits and expectations of our bodies, ultimately contributing to a discussion that will determine new categories of disease. It's about what we value, and what we class as normal. And it's about the voice we use to communicate those positions.

The outcomes of these decisions aren't trivial. Even for those of us who aren't personally unwell, our line between health and disease impacts on many facets of our lives.

Our definition of disease tells us what we should fund for research, welfare and insurance. It determines who qualifies for help and who doesn't. It's about access to medication, to assistance, to information. It's the difference between privilege and rights, entitlement and pleasure.

Our definition of disease tells us who to treat as broken. It's a statement of what defines 'normal', and therefore 'valuable'. It sets both a standard we should strive for, and a limit we can't move beyond. It says help is there for those who don't meet our standards, but not necessarily for those who don't meet their own.

Our definition of disease tells us who has power and agency. It tells us who is responsible for a human body in life and in death. It defines lines of control and impetus for action.

Our definition of disease tells us who is guilty and who is innocent, who deserves punishment and retribution and who deserves help and rehabilitation. It tells us who should be ashamed and who should be pitied.

For centuries, the defining characteristics of disease have evolved to reflect a neat dichotomy, where disease's discrete nature has been shaped by the merging of cultures, by new discoveries in biology and by advances in medical technology. Those characteristics will continue to evolve. How we see disease in the future – if it persists as a concept at all – will be significantly different to how we see it now, just as it has been significantly different in the past.

Signs hinting at what disease might look like one day aren't hard to find. This time we have 'big data' and advancing information technology to thank for the changes. In 1980 you needed a hard disk the size of a fridge to store a single gigabyte of data. Not even half a century later, we can store hundreds of times that amount of information in the palm of our hand. And that's a good thing, because in 2017 we generated 2.5 quintillion bytes of data – every day.[43] That's 25 followed by seventeen zeros. And that figure is growing; by the time you read this, that number will be obsolete, replaced by an even bigger figure.

Put into historical terms, some 90 per cent of humanity's total information output – whether it's fan fiction, fake news, Instagram photos, impressionist paintings, Greek sculptures, philosophical musings, scientific research, maps of distant lands or measurements of the hydrogen fingerprints of distant stars – has been generated in the past two years. We don't even read information anymore: we mine it, as if it's our mental sediment compressed into evidence of human nature.

This reaping and storing of information is profoundly changing how we do medical science. We can plough through oceans of numbers using algorithms that search out patterns, looking for hints of growing epidemics in our Google search results, clues about mental health in our Twitter feeds, strategies on improving nutrition on our Pinterest walls. We can aggregate a history of research gathered from massive databases and evaluate them over and over again, refining our previous guesses on which genes or synapses are responsible for making us unwell, to paint a richly complex picture of our physiological diversity.

Growing pools of data expose health as a fractal mosaic, where the closer we look, the harder it gets to separate our bodies into neat categories of normal versus broken. Take cancer, for example. Once, it was a disease defined as a tangible lump of tissue that strangled the life out of our vital organs. With advances in microscopy and better tissue staining, we've come to see it as a diversity of illnesses marked by differences in cells and aberrations in biochemical control processes. Discoveries in genetics have expanded our knowledge of how cells grow and evolve, and how they change and evade detection in the body. Further finds in epigenetics add even more detail: how a cell can break free of its biochemical shackles, for instance, and go rogue under environmental stresses. Each new layer of knowledge provides an exponential surge in data, demanding ever more complex models to map and understand the interactions between cancerous cells and the rest of our bodies. Far from being a simple disease, it's clear that cancer is almost as diverse as the individuals it affects.

To ease suffering, oncologists are increasingly employing a personalised approach to medicine, using pinpoint-accurate genetic screens and sensitive bioassays to tailor bespoke regiments of cellular toxins, all working in combination to suit individual needs. A radical new method already being tested is growing biopsied tissue samples in the lab as 'organoids', then treating them as mini-tumours – an in-vitro proxy on which we can test out recipes of treatments before risking them on a patient.[44] Soon we could be generating epigenetic blockers or strings of nucleic acid that interfere with the genetic elements in your cancerous cells, or maybe even devise and print proteins with precisely the right kinks to fit your particular molecular physiology.

It isn't a stretch to imagine this approach being applied to other medical fields. The top ten highest-grossing drugs in the United States help as few as one in twenty-five people who take them. The statins we take to control our cholesterol levels might be relatively useless for 98 per cent of people who use them.[45] That doesn't make these medications poor choices; on balance, they save lives. But they're also based on generalisations, often informed by a statistical normality set by the race, socioeconomic status, health, weight, gender and age of the volunteers who trial medications.

Medicine has worked this way in part thanks to a history of assumptions about health and disease. But it's mostly a constraint of the data: it's far easier to collect a large sample size of people and make a broad evaluation of the efficacy of new treatments than it is to collect mountains of data on the nature of each individual. Modelling a population based on generalisations and broad categories is far more feasible than accurately modelling the diversity within it.

This is slowly changing. In April 2003 the Human Genome Project delivered its goal of sequencing most of what we call the *euchromatic* parts of our genome, which make up around 92 per cent of our code.[46] In the years since, we've used this data to build libraries that can better comprehend not just our unique genomes but also the individual way they operate and the way our environment interacts with them. In January 2015 President Barack Obama announced funding of US$215 million for the National Institutes of Health to collect genetic information from 1 million people.[47] Called the All of Us research program, its goal is to build a foundation for improving precision medicine.

Imagine matching psychiatric treatments to individual balances of neurotransmitters or neurological anatomy. Pulling mental health symptoms that weren't picked up by physicians in electronic records is already becoming a reality, with studies making progress on methods that could tease out concerning signs of illness that would help inform better treatments. Combined with DNA banks, this is shaping as a powerful tool with which we can understand the complexity of health based on individual symptoms, without needing to make sense of it in terms of overarching diseases.[48]

We're beginning to fathom the fingerprint of microflora in each of our guts, and how they're responsible for our health and wellbeing. Right now, we do our best to maintain their health with diet, probiotics and – yes, it's a thing – faecal transplants. In time, we might have a comprehensive way to measure and affect the make-up of our digestive system's microbes in such a way that we might prevent ailments long before they arise, all by fiddling with our germs.

With all of this on the horizon, disease starts to look like an old-fashioned idea. The focus of medicine can return to the idea of individual suffering and its aetiology, leapfrogging the moral question of whether you belong in a sanctioned category of health. It can once again simply ask: 'How can we help you feel better?'

Rather than questioning whether obesity is a disease, we'll accept the myriad ways people suffer or enjoy their bodies, and offer opportunities to ease suffering based on their specific biology and circumstances. Instead of asking whether ADHD is a true disorder, we might eventually come to see pharmaceuticals as an opportunity for all of us to responsibly shape our biology to suit our desires. It won't matter whether it's an addiction or simply an obsession – if it makes life for us and our loved ones unbearable, we'll find the right approach to getting life back on track. Growing organoids, running genetic screens and neurological scans, creating bespoke drug regiments – this, I hope, is the medicine of the future.

Idealistic? Absolutely. Lots of factors on multiple levels of society would need to change before such a world would be possible, and many of them would be stubborn to shift. For example, just who owns our information? Generalised statistics belong to a population, but the data describing your own health is like our health itself – it exists as some sort of superposition, inside both us as individuals and collectively as the community. To get to a future where we personalise our health down to such extraordinary detail, we need to overcome problems of data privacy and ownership.

A post-disease world would be economically, morally and politically different to the one we have now. It would demand differences in how physicians consult and how we invest in health care. It would mean stopping asking things like 'Is X a disease?' and 'Are we getting our money's worth?'

In any case, our future is going to be different anyway. Our notion of health and disease won't be the same as it is today, just as it isn't the same today as it was a century ago. We can expect there to be new versions of the *ICD* and *DSM* for decades to come, of course, and we can expect that there will always be an authority listing the consequences of concerning high-risk pathogens and biological variants, according to the latest opinions and statistics. These authorities of health will undoubtedly evolve, though.

In many aspects, this shift won't be all that far into the future. It's already present in patches: for every physician who appears to be unduly focused with tunnel-vision on a diagnosis, there are many who appreciate the basic need to ease their patients' suffering by taking into account their specific needs. For each doctor who feels pressed by time to gamble the best treatment based on their medical program's advice, there will be others who take more time to listen to a patient's concerns. It's present in teachers who see each student as the sum of factors that contribute to their learning, and who work with parents to deliver adequate learning support. It's here in parents who don't wait for a diagnosis to tell them how to best help their child get the most out of life, and yet who still value a diagnosis as a stepping stone to opportunity.

This future is the result of accepting that Mother Nature doesn't have a moral position on our health or biological abilities. There is no divine goal, no grand plan, no universal archetype of human form. To borrow internet slang, she DNGAF.

Medicine isn't about fixing something that is broken. It's about alleviating hurt and suffering. And perhaps the best way we can go about that is to dream of a future in which we've left today's concept of disease behind us.

# ACKNOWLEDGEMENTS

*Unwell* had three moments of conception.

The first was in 2012, soon after the release of *Tribal Science*. I was midway through doing a bunch of postgraduate courses on the culture of health and medicine, pulling together readings for a paper on crime and disease and conducting interviews on how we define both. Initially, I intended to flesh out a few reports I'd written on the history and culture of mental health, but drawing that line seemed to repeat the same mistake of treating the human mind as fundamentally different to other functions of the body. I could see a deeper story there, but couldn't quite work out how to approach it.

Three years later, I returned to the project after meeting the American anti-poverty campaigner Linda Tirado over a scotch, prior to her appearance on ABC TV's *Q&A* program. Her passion for tackling misinformation and biases against society's poor inspired me to return to my own efforts to better understand the conversation about society's outcast and unwell. I went back to the old documents I'd started and fleshed out the bones of a book on how the boundaries of disease and medicine are defined by an authority. For her prompting, I'm forever in Linda's debt.

What you've read here isn't quite what I wrote then, though. The final version is far more personal than I ever imagined it might be. As a science writer, weaving yourself into a narrative feels like swimming against a tide of instinctive objectivity. Add to that the fact I'm not a doctor, and don't identify myself as unwell in any way, and I felt – and still feel – somewhat like a tourist. This never felt like my own story to tell, and yet here we are.

The final conception came about following the advice of my agent, Pippa Masson, and her team at Curtis Brown, who encouraged me not to fear adding my own voice and my own stories. I started afresh, forcing myself to examine my privileged life as it is influenced by the definition of disease, and discovered that leaving my own narratives out of my discussion of health and disease would have been telling only half the story. Thank you, Pippa and Caitlan Cooper-Trent, for your suggestions and persistence.

That said, please don't think I have the first clue about what it's like to be affected by a debilitating illness. I have never had my health seriously questioned, and aside from uncomfortable cycles of depression since my teens, I have never questioned the state of my own body or mind. So it's been a struggle to know how to best blend my passion for the history of disease with the lived experiences of others, without telling their stories for them. I'm all too aware that this lack of experience has left me prone to making all kinds of mistakes in communicating the social history and research behind disease, and I hope you'll forgive any errors that have slipped through.

Fortunately, I have had many wonderful people to keep me in check.

Paige Wilcox and Liz Duck-Chong were invaluable in their advice on appropriate language when discussing gender and sexuality. I could not have done without Ada Quinn, who not only took the time to provide corrections on sensitive terms, but did so sympathetically, through the eyes of a scientist. I am deeply indebted to you all, and while I know it's still not perfect, I promise I did the best I could.

From the day we sat in a Newtown bar discussing medieval food culture, Emma Groucutt has been a constant force in making this book as accurate as possible. Her research skills and corrections regarding the history of sexuality, bathing and diet, not to mention her regular suggestions of reputable texts (a pile that continues to grow), were a vital resource for which I am eternally grateful.

I must also thank Jack Neens for our conversations on current trends in psychology and the history of CFS; Katie Hail-Jares for her expertise on the hazy boundaries between criminology and biology; James Clarke for our chats on addiction, and how the Australian NDIS defines disability; and Ash Werchanowski for bringing me down to earth on current medical culture (and the apocalyptic terror of the P-zombie). You all gave me a solid wall to bounce my understanding off when I needed it.

Eliza Bortolotti has been a rock throughout the entire process, reading and rereading drafts, giving me advice on wording, hunting down medical papers and generally putting up with my endless waffling over the years. With doctors like you in our future, I have no doubt we'll all be in good hands.

Of course, my wife, Liz, has been a robust support as always. For more than sandwiches. We've gone through much over the years, but her patient exploration of her own and our son Asher's challenges has modelled for me how to strike the right balance in living with what the world sees as a disorder.

And then there is Liz Caplice. I told you I'd make you immortal, for a little while longer. Not that you needed it – your wisdom shines bright in so many people.

*Unwell* wouldn't be what it is if not for Julian Welch's professional editing, which cleared the pathways in what was effectively an overgrown garden, and helped me avoid some embarrassing errors along the way. I'm so appreciative of that eagle eye. Lastly, a big thanks to Alexandra Payne and UQP for taking this project out of my head and putting it into yours.

# NOTES

## PART I BROKEN

1. Barkley, R. (2011) When the Diagnosis Is A.D.H.D. *The New York Times*, 15 February 2011.
2. Hilts, P.J. (1993) Gulf War Syndrome: Is It a Real Disease? *The New York Times*, 23 November 1993.
3. CJZ, *The Obesity Myth*. SBS Television, 4 September 2017.
4. Boycott, R. & Lawson, D. (2007) Is Addiction an Illness or a Weakness? *Daily Mail Online*, 31 March 2007.
5. LaMattina, J. (2013) Restless Legs Syndrome – A Pharma Creation or a Real Health Problem? *Forbes*, 22 June 2013.
6. Macmillan, A. (2017) Is Sex Addiction Real? Here's What Experts Say. *Time*, 10 November 2017.
7. Hastings, R. (1898) Noise as a Factor in the Causation of Disease. *JAMA*, vol. 31 (26), pp. 1522–1524.
8. Hofer, J. (1934) Medical Dissertation on Nostalgia by Johannes Hofer, 1688, translated by Anspach, Carolyn Kiser. *Bulletin of the Institute of History of Medicine*, vol. 2, pp. 376–391.
9. Banks, J. (1820) *The Endeavour Journal of Sir Joseph Banks*. Gutenberg.org, November 2005.
10. As in Roth, M.S. (1991) Dying of the Past: Medical Studies of Nostalgia in Nineteenth-century France. *History and Memory*, vol. 3 (1), pp. 5–29.
11. Avicenna, The Canon of Medicine, in Bakhtiar, L. (1999) *Avicenna Canon of Medicine Volume 1*. Kazi Publications.
12. Councilman, W.T. (1913) *Disease and Its Causes*. Henry Holt and Company.
13. World Health Organization (1946) *Preamble to the Constitution of the World Health Organization*. WHO, New York.
14. Rosenberg, C. & Golden, J. (1992) *Framing Disease: Studies in Cultural History*. Rutgers University Press.
15. van Foreest, P. (1623) *Observationem et Curationem Medicinalium ac Chirurgicarum Opera Omnia*.
16. Granville, J.M. (1883) *Nerve-vibration and Excitation as Agents in the Treatment of Functional Disorder and Organic Disease*. J&A Churchill, via archive.org.
17. Sydenham, T. (1734) *The Whole Works of Thomas Sydenham*. W. Feales.
18. As in King, H. (1996) Green Sickness: Hippocrates, Galen and the Origins of the 'Disease of Virgins'. *International Journal of the Classical Tradition*, vol. 2 (3), pp. 372–387.

19. Lloyd, E. (2005) *The Case of the Female Orgasm*. Harvard University Press.
20. Hogg, J. (1830) *The Raid of the Kerrs*. Clankerr.co.uk.
21. Mandal, M. & Dutta, T. (2001) Left Handedness: Facts and Figures across Cultures. *Psychology and Developing Countries*, vol. 13 (2), pp. 173–191; McManus, I. (2009) The History and Geography of Human Handedness. In Sommer, I.E.C. & Kahn, R.S. (eds), *Language Lateralization and Psychosis*. Cambridge University Press.
22. Bryden, M.P., Ardila, A. & Ardila, O. (1993) Handedness in Native Amazonians. *Neuropsychologia*, vol. 31 (3), pp. 301–308.
23. Ling-fu, M. (2007) The Rate of Handedness Conversion and Related Factors in Left-handed Children. *Laterality: Asymmetries of Body, Brain and Cognition*, vol. 12 (2), pp. 131–138.
24. Siebner, H., Limmer, C., Peinemann, A., Drzezga, A., Bloem, B., Schwaiger, W. & Conrad, B. (2002) Long-term Consequences of Switching Handedness: A Positron Emission Tomography Study on Handwriting in 'Converted' Left-handers. *Journal of Neuroscience*, vol. 22 (7), pp. 2816–2825.
25. Hopkins, W.D., Wesley, M.J., Izard, M.K., Hook, M. & Schapiro, S.J. (2007) Chimpanzees (Pan troglodytes) Are Predominantly Right-handed: Replication in Three Populations of Apes. *Behavioural Neuroscience*, vol. 118 (3), pp. 659–663.
26. Brown, C. & Magat, M. (2011) Cerebral Lateralization Determines Hand Preferences in Australian Parrots. *Biology Letters*, vol. 7, pp. 496–498.
27. Arning, L. et al. (2015) Handedness and the X Chromosome: The Role of Androgen Receptor CAG-repeat Length. *Scientific Reports*, vol. 5 (8325).
28. Blau, T.H. (1977) Torque and Schizophrenic Vulnerability: As the World Turns. *American Psychology*, vol. 32 (12), pp. 997–1005.
29. Geschwind, N. (1982) Left-handedness: Association with Immune Disease, Migraine, and Developmental Learning Disorder. *PNAS*, vol. 79 (16), pp. 5067–5100.
30. Coren, S. & Halpern, D. (1991) Left-handedness: A Marker for Decreased Survival Fitness. *Psychological Bulletin*, vol. 109, pp. 90–106.
31. Annett, M. (1993) The Fallacy of the Argument for Reduced Longevity in Left-handers. *Perceptual and Motor Skills*, vol. 76, pp. 295–298.
32. Kreisfeld, R. & Newson, R. (2006) *Hip Fracture Injuries*. AIHW National Injury Surveillance Unit.
33. Australian Institute of Health and Welfare (2008) *A Picture of Osteoporosis in Australia*. Arthritis series no. 6, cat. no. PHE 99. Canberra: AIHW.
34. Newton-John, H.F. & Morgan, D.B. (1968) Osteoporosis: Disease or Senescence? *The Lancet*, vol. 291 (7536), pp. 232–233.
35. World Health Organization (1994) *Assessment of Fracture Risk and Its Application to Screening for Postmenopausal Osteoporosis: Report of a WHO Study Group*. Geneva: World Health Organization.
36. World Health Organization (2015) Globally, an Estimated Two-thirds of the Population Under 50 Are Infected with Herpes Simplex Virus Type 1. News release: www.who.int/mediacentre/news/releases/2015/herpes/en.
37. *Newsweek* staff (1990) *Chronic Fatigue Syndrome*. *Newsweek*, 11 November 1990.
38. Bates, C. (2007) Doctors Are Ordered to Take 'Yuppie Flu' Seriously. *Daily Mail*, 23 August 2007.

39. Liddle, R. (2012) 'Pretend Disabled' Really ARE Sick. *The Sun*, 26 January 2012.
40. Nagy-Szakal, D. et al. (2017) Fecal Metagenomic Profiles in Subgroups of Patients with Myalgic Encephalomyelitis/Chronic Fatigue Syndrome. *Microbiome*, vol. 5 (44), doi.org/10.1186/s40168-017-0261-y.
41. Tomas, C., Brown, A., Strassheim, V., Elson, J., Newton, J. & Manning, P. (2017) Cellular Bioenergetics Is Impaired in Patients with Chronic Fatigue Syndrome. *PLOS One*, vol. 12 (10), doi.org/10.1371/journal.pone.0186802.
42. Response to 'How do you rate the management of your ME/CFS that is provided by your usual NHS GP?' in the UK ME Association Archive of MEA Surveys, N=488, date ending 10 March 2016, www.meassociation.org.uk/archive-of-mea-surveys, last accessed 17 February 2018.
43. Johnston, S., Brenu, E., Staines, D. & Marshall-Gradisnik, S. (2013) The Prevalence of Chronic Fatigue Syndrome/Myalgic Encephalomyelitis: A Meta-analysis. *Clinical Epidemiology*, vol. 5, pp. 105–110.
44. Collin, S., Bakken, I., Nazareth, I., Crawley, E. & White, P.D. (2017) Trends in the Incidence of Chronic Fatigue Syndrome and Fibromyalgia in the UK, 2001–2013: A Clinical Practice Research Datalink Study. *Journal of the Royal Society of Medicine*, vol. 110 (6), pp. 231–244.
45. Young, A. (1982) The Anthropologies of Illness and Sickness. *Annual Review of Anthropology*, vol. 11, pp. 257–285.
46. Gander, K. (2014) Why Do Some Forms of Cancer Receive More Research Funding Than Others? *The Independent*, 2 October 2014.
47. Szasz, T. (2011) [1961], *The Myth of Mental Illness: Foundations of a Theory of Personal Conduct*. Harper & Row.
48. Cartwright, S.A. (1851) Excerpt from *Diseases and Peculiarities of the Negro Race*. De Bow's Review Southern and Western States, vol. XI, New Orleans, via www.pbs.org.
49. Anderson, M. & Harkness, S. (2017) When Do Biological Attributions of Mental Illness Reduce Stigma? Using Qualitative Comparative Analysis to Contextualize Attributions. *Society and Mental Health*, doi.org/10.1177/2156869317733514.

## PART II INFECTED

1. Herlihy, D. (1997) *The Black Death and the Transformation of the West*. Harvard University Press.
2. Allen, B. (2015) *The Effects of Infectious Disease on Napoleon's Russian Campaign*. Pickle Partners Publishing.
3. Morens, D.M. & Taubenberger, J.K. (2006) 1918 Influenza: The Mother of All Pandemics. *Emerging Infectious Diseases*, vol. 12 (1), pp. 15–22.
4. Jouanna, J. (2012) Air, Miasma, and Contagion in the Time of Hippocrates and the Survival of Miasmas in Post-Hippocratic Medicine (Rufus of Ephesus, Galen and Palladius). In Jouanna, J. (ed.), *Greek Medicine from Hippocrates to Galen: Select Papers*. Brill, pp. 119–136.
5. Rowland, I.D. & Noble Howe, T. (ed. and trans.) (2001) *Vitruvius: 'Ten Books on Architecture'*. Cambridge University Press.
6. Aberth, J. (2012) *An Environmental History of the Middle Ages: The Crucible of Nature*. Routledge.

7. General Board of Health (1849) *Report on Quarantine*. London: William Clowes & Sons, via archive.org.
8. Fracastoro, G. (1530) *Syphilis, or a Poetical History of the French Disease*. Translated by N. Tate, via quod.lib.umich.edu.
9. Fletcher, J. (2011) *A Study of the Life and Works of Athanasius Kircher, 'Germanus Incredibilis'*. Brill.
10. Croft, W. (2006) *Under the Microscope: A Brief History of Microscopy*. World Scientific Publishing Co.
11. Barnett, J. (2003) Beginnings of Microbiology and Biochemistry: The Contribution of Yeast Research. *Microbiology*, vol. 149, pp. 557–567.
12. Nuland, S. (2004) *The Doctors' Plague Germs, Childbed Fever, and the Strange Story of Ignac Semmelweis*. W.W. Norton & Co.
13. Tweedie, A. (1841) *The Library of Medicine, Midwifery, Vol. VI*. London: Whittaker and Co., p. 292.
14. As referenced by Pasteur, L. (1864) On Spontaneous Generation. *Revue des cours scientifics*, vol. 1, pp. 257–264, citing 'The works of Jan Baptiste van Helmont dealing with the principles of medicine and physics for the insured cure of diseases' (1670), translated by Jean Le Conte.
15. Birch, T. (ed.) (1772) *The Works of the Honourable Robert Boyle*. London. In Clericuzio, A. (1990) A Redefinition of Boyle's Chemistry and Corpuscular Philosophy. *Annals of Science*, vol. 47, pp. 561–589.
16. Evans, A. (1976) Causation and Disease: The Henle-Koch Postulates Revisited. *Yale Journal of Biology and Medicine*, vol. 49 (2), pp. 175–195.
17. Loeffler, F. & Frosch, P. (1898) Reports from the Commission on the Study of Foot-and-mouth Disease at the Institute for Infectious Diseases in Berlin for Bacteriology. *Central Journal of Parasitic and Infectious Diseases*, vol. 23, pp. 371–391.
18. Carpenter, K. (2000) *Beriberi, White Rice and Vitamin B*. University of California Press.
19. Young, A. (1927) Lord Lister's Life and Work. *Janus*, vol. 31. In Gaw, J. (1999) 'A Time to Heal': The Diffusion of Listerism in Victorian Britain. *American Philosophical Society*, vol. 89 (1), p. 8.
20. Chaloner, E.J., Flora, H.S. & Ham, R.J. (2001) Amputations at the London Hospital 1852–1857. *Journal of the Royal Society of Medicine*, vol. 94 (8), pp. 409–412.
21. Nightingale, F. (1969) [1858] *Notes on Nursing: What It Is and What It Is Not*. Dover.
22. Lister, J. (1867) On the Antiseptic Principle in the Practice of Surgery. *The Lancet*, vol. 93, pp. 353–356.
23. Friday, J. (2009) A Microscopic Incident in a Monumental Struggle: Huxley and Antibiosis in 1875. *The British Journal for the History of Science*, vol. 7 (1).
24. Center for Disease Dynamics, Economics & Policy (2015) *State of the World's Antibiotics, 2015*. Washington DC: CDDEP.
25. Dyson, F. (2000) Gravity Is Cool, or, Why Our Universe Is Hospitable to Life. Oppenheimer lecture, University of California Berkeley, California, www.ocf.berkeley.edu/~fricke/dyson.html.
26. I highly recommend reading Ed Yong's amazing book, *I Contain Multitudes: The Microbes Within Us and a Grander View of Life* (2016), for a deeper look at our intimate relationship with microbes.

27. Bolinger, R. et al. (2007) Biofilms in the Large Bowel Suggest an Apparent Function of the Human Vermiform Appendix. *Journal of Theoretical Biology*, doi:10.1016/j.jtbi.2007.08.032.
28. Hacker, J. et al. (1997) Pathogenecity Islands of Virulent Bacteria: Structure, Function and Impact on Microbial Evolution. *Molecular Microbiology*, vol. 23 (6), pp. 1089–1097, doi:10.1046/j.1365-2958.
29. Marshall, B. & Warren, J. (1984) Unidentified Curved Bacilli in the Stomach of Patients with Gastritis and Peptic Ulceration. *The Lancet*, doi:10.1016/S0140-6736(84)91816-6.
30. Morais, S. et al. (2016) Trends in Gastric Cancer Mortality and in the Prevalence of Helicobacter Pylori Infection in Portugal. *European Journal of Cancer Prevention*, vol. 25 (4), pp. 275–281.
31. Xie, F. et al. (2013) Helicobacter Pylori Infection and Esophageal Cancer Risk: An Updated Meta-analysis. *World Journal of Gastroenterology*, vol. 19 (36), pp. 6098–6107.
32. Falk, G. (2001) The Possible Role of Helicobacter Pylori in GERD. *Seminars in Gastrointestinal Disease*, vol. 12 (3), pp. 186–195.
33. Arnold, I.C. et al. (2011) Helicobacter Pylori Infection Prevents Allergic Asthma in Mouse Models Through the Induction of Regulatory T Cells. *Journal of Clinical Investigation*, vol. 121 (8), pp. 3088–3093.
34. Chen, Y. et al. (2013) Association Between Helicobacter Pylori and Mortality in the NHANES III Study. *Gut*, vol. 62 (9), pp. 1262–1269.
35. Stokholm, J. et al. (2017) Cat Exposure in Early Life Decreases Asthma Risk from the 17q21 High-risk Variant. *Journal of Allergy and Clinical Immunology*, doi.org/10.1016/j.jaci.2017.07.044.
36. Tun, H. et al. (2017) Exposure to Household Furry Pets Influences the Gut Microbiota of Infants at 3–4 Months Following Various Birth Scenarios. *Microbiome*, vol. 5 (40).
37. Holbriech, M. et al. (2012) Amish Children Living in Northern Indiana Have a Very Low Prevalence of Allergic Sensitization. *The Journal of Allergy and Clinical Immunology*, vol. 129 (6), pp. 1671–1673.
38. Shuman, R.M., Leech, R.W. & Alvord Jr., E.C. (1975) Neurotoxicity of Hexachlorophene in Humans II. A Clinicopathological Study of 46 Premature Infants. *Archives of Neurology*, vol. 32 (5), pp. 320–325.
39. Food and Drug Administration (2016) FDA Issues Final Rule on Safety and Effectiveness of Antibacterial Soaps. www.fda.gov/NewsEvents/Newsroom/PressAnnouncements/ucm517478.htm.
40. Yueh., M. et al. (2014) The Commonly Used Antimicrobial Additive Triclosan Is a Liver Tumor Promoter. *PNAS*, vol. 111 (48), doi:10.1073/pnas.1419119111.
41. Dhillon, G.S. et al. (2015) Triclosan: Current Status, Occurrence, Environmental Risks and Bioaccumulation Potential. *International Journal of Environmental Research and Public Health*, vol. 12 (5), pp. 5657–5684.
42. Center for Disease Dynamics, Economics & Policy (2015) *State of the World's Antibiotics, 2015*. Washington DC: CDDEP.
43. Centers for Disease Control and Prevention (2013) *Antibiotic Resistance Threats in the United States, 2013*. Report, www.cdc.gov/drugresistance/threat-report-2013.
44. Chan, M. (2014) WHO Director-General Addresses the Regional Committee for Africa. www.who.int/dg/speeches/2014/regional-committee-africa/en.

## PART III INSANE

1. Stanton, M.D. (1972) Drug Use in Vietnam: A Survey Among Army Personnel in the Two Northern Corps. *Archives of General Psychiatry*, vol. 26.
2. Robins, L.N., Helzer, J.E. & Davis, D.H. (1975) Narcotic Use in Southeast Asia and Afterward: An Interview Study of 898 Vietnam Returnees. *Archives of General Psychiatry*, vol. 32 (8), pp. 955–961.
3. Smyth, B.P., Barry, J., Keenan, E. & Ducray, K. (2010) Lapse and Relapse Following Inpatient Treatment of Opiate Dependence. *Irish Medical Journal*, vol. 103 (6), pp. 176–179.
4. Alexander, B.K., Coambs, R.B. & Hadaway, P.F. (1978) The Effect of Housing and Gender on Morphine Self-administration in Rats. *Psychopharmacology*, vol. 58, pp. 175–179.
5. Alexander, B.K. (2010) *Addiction: The View from Rat Park*. www.brucekalexander.com.
6. Petrie, B.F. (1996) Environment Is Not the Most Important Variable in Determining Oral Morphine Consumption in Wistar Rats. *Psychological Reports*, vol. 78 (2), pp. 391–400.
7. Rudd, R.A., Seth, P., David, F. & Scholl, L. (2016) Increases in Drug and Opioid-involved Overdose Deaths – United States, 2010–2015. *MMWR Morb Mortal Wkly*, vol. 65, pp. 1445–1452.
8. Vowles, K.E., McEntee, M.L., Julnes, P.S., Frohe, T., Ney, J.P. & van der Goes, D.N. (2015) Rates of Opioid Misuse, Abuse, and Addiction in Chronic Pain: A Systematic Review and Data Synthesis. *Pain*, vol. 156 (4), pp. 569–576.
9. National Institute of Health. *Overdose Death Rates*, www.drugabuse.gov/related-topics/trends-statistics/overdose-death-rates. Last accessed February 2018.
10. Nutt, D., King, L.A. & Phillips, L.D. (2010) Drug Harms in the UK: A Multicriteria Decision Analysis. *The Lancet*, vol. 376 (9752), pp. 1558–1565.
11. Sharma, N. (2015) This Neuroscientist Argues that Addiction Is Not a Disease and Rehab Is Bullshit. *Vice Magazine*, 5 December 2015.
12. Ouellette, J. (2016) Addiction Could Be More Like a Learning Disorder than a Disease. *Gizmodo*, 19 April 2016.
13. Manjila, S. et al. (2008) Modern Psychosurgery Before Egas Moniz: A Tribute to Gottlieb Burckhardt. *Neurosurgery Focus*, vol. 25 (1).
14. Lombroso-Ferrero, G. (2009) Criminal Man, According to the Classification of Cesare Lombroso. Via Project Gutenberg.
15. Lombroso, C. (1903) Left-handedness and Left-sidedness. *The North American Review*, vol. 177 (562), pp. 440–444.
16. Gross, D. & Schäfer, G. (2011) Egas Moniz (1874–1955) and the 'Invention' of Modern Psychosurgery: A Historical and Ethical Reanalysis Under Special Consideration of Portuguese Original Sources. *Neurosurgical Focus*, vol. 30 (2), doi:10.3171/2010.10.FOCUS10214.
17. Holland, R., Kopel, D., Carmel, P.W. & Prestigiacomo, C.J. (2017) Topectomy versus Leukotomy: J. Lawrence Pool's Contribution to Psychosurgery. *Neurosurgical Focus*, vol. 43 (3).
18. Rendell, J., Huss, M. & Jensen, M. (2010) Expert Testimony and the Effects of a Biological Approach, Psychopathy and Juror Attitudes in Cases of Insanity. *Behavioural Sciences and the Law*, vol. 28, pp. 411–425.

19. Townsend, M. (2012) Breivik Verdict: Norwegian Extremist Declared Sane and Sentenced to 21 Years. *The Guardian*, 24 August 2012.
20. BBC News (2011) Norway Massacre: Breivik Declared Insane. *BBC News*, 29 November 2011.
21. Townsend, M. (2012) Anders Breivik Should Be Declared Sane, Majority of Norwegians Believe. *The Guardian*, 24 April 2012.
22. Reuters staff (2012) Norway Killer – 'Insane' Diagnosis 'Worse than Death'. Reuters, 4 April 2012.
23. Oshisanya, O. (2018) *An Almanac of Contemporary and Comparative Judicial Restatements* (ACCJR Supp. ii Public Law), Almanac Foundation.
24. *Australian Criminal Code Act 1995*, Division 7 – Circumstances involving lack of capacity, 7.3(9), compiled 13 December 2017.
25. Fox, R. (1988) The Killings of Bobby Veen: The High Court on Proportion in Sentencing. *Criminal Law Journal*, vol. 12, 339–352.
26. Freeman, K. (1998) Mental Health and the Criminal Justice System. *Contemporary Issues in Crime and Justice*, no. 38.
27. Human Rights and Equal Opportunity Commission (1993) *Human Rights and Mental Illness: Report of the National Inquiry into the Human Rights of People with Mental Illness* (the 'Burdekin Report'). AGPS.
28. Vilares, I. et al. (2017) Predicting the Knowledge: Recklessness Distinction in the Human Brain. *Proceedings of the National Academy of Sciences*, vol. 114 (12), pp. 3222–3227.
29. Aharoni, E. et al. (2013) Neuroprediction of Future Rearrest. *PNAS*, vol. 110 (15), doi:10.1073/pnas.1219302110.
30. Darby, R., Horn, A., Cushman, F. & Fox, M. (2018) Lesion Network Localization of Criminal Behavior. *PNAS*, vol 115 (3), pp. 601–606.
31. Tiihonen, J. et al. (2014) Genetic Background of Extreme Violent Behaviour. *Molecular Psychiatry*, vol. 20 (6), pp. 786–792.
32. Hook, G.R. (2009) 'Warrior Genes' and the Disease of Being Māori. *MAI Review*.
33. Burke, G. (2017) Gay Panic Defence Scrapped by Queensland Parliament after Lobbying by Catholic Priest. *ABC News*, 22 March 2017.
34. As quoted in Crompton, L. (2003) *Homosexuality and Civilization*. Belknap Press.
35. Aristotle (2011) *Problems*, Book IV, Loeb Classical Library. Harvard University Press.
36. LeVay, S. (1991) A Difference in Hypothalamic Structure Between Heterosexual and Homosexual Men. *Science*, vol. 253 (5023), pp. 1034–1037.
37. Hu, S. et al. (1995) Linkage Between Sexual Orientation and Chromosome Xq28 in Males but Not in Females. *Nature Genetics*, vol. 1 (3), pp. 248–256.
38. Sanders, A.R. et al. (2015) Genome-wide Scan Demonstrates Significant Linkage for Male Sexual Orientation. *Psychological Medicine*, vol. 45 (7), pp. 1379–1388.
39. Rice, G., Anderson, C., Risch, N. & Ebers, G. (1999) Male Homosexuality: Absence of Linkage to Microsatellite Markers at Xq28. *Science*, vol. 284 (5414), pp. 665–667.
40. Bearman, P.S. & Bruckner, H. (2002) Opposite-sex Twins and Adolescent Same Sex Attraction. *American Journal of Sociology*, vol. 107 (5), pp. 1179–1205; Whitlam, F.L., Diamond, M. & Martin, J. (1991) Homosexual Orientation in Twins: A Report on 61 Pairs and Three Triplet Sets. *Archives of Sexual Behavior*, vol. 22 (3), pp. 187–206.

41. Reardon, S. (2015) Epigenetic 'Tags' Linked to Homosexuality in Men. *Nature*, doi:10.1038/nature.2015.18530.
42. Copen, E., Chandra., A. & Febo-Vazquez, I. (2016) *Sexual Behavior, Sexual Attraction, and Sexual Orientation Among Adults Aged 18–44 in the United States: Data From the 2011–2013 National Survey of Family Growth*. National Health Statistics Reports, No. 88.
43. Tomazin, F., Grace, R. & Koziol, M. (2018) Gay 'conversion' off Liberal agenda after party president pulls pin. *The Age*, 16 April 2018.
44. Fayyad, J., De Graaf, R., Kessler, R. et al. (2007) Cross-national Prevalence and Correlates of Adult Attention-deficit Hyperactivity Disorder. *British Journal of Psychiatry*, vol. 190, pp. 402–409.

## PART IV ABNORMAL

1. World Health Organization (2010) *Neonatal and Child Male Circumcision: A Global Review*. UNAIDS.
2. The numbers are pretty hard to pin down with much accuracy, and depend on the methods used and the years covered. This rough breakdown comes from Morris, B.J. et al. (2016) Estimation of Country-specific and Global Prevalence of Male Circumcision. *Population Health Metrics*, vol. 14 (4).
3. Pfuntner, A., Wier, L.M. & Stocks, C. (2011) *Most Frequent Procedures Performed in U.S. Hospitals*. Healthcare Cost and Utilization Project (HCUP) Statistical Brief.
4. Rediger, C. (2013) Parents' Rationale for Male Circumcision. *Canadian Family Physician*, vol. 59 (2).
5. Xu, B. & Goldman, H. (2008) Newborn Circumcision in Victoria, Australia: Reasons and Parental Attitudes. *ANZ Journal of Surgery*, vol. 78 (11), pp. 1019–1022.
6. Tiemstra, J. (1999) Factors Affecting the Circumcision Decision. *Journal of the American Board of Family Medicine*, vol. 12 (1), pp. 16–20.
7. Sardi, L. & Livingston, K. (2015) Parental Decision Making in Male Circumcision. *MCN, the American Journal of Maternal Child Nursing*, vol. 40 (2), pp. 110–115.
8. US Centers for Disease Control and Prevention (2014) *Recommendations for Providers Counseling Male Patients and Parents Regarding Male Circumcision and the Prevention of HIV Infection, STIs, and Other Health Outcomes*.
9. Barnholtz-Sloan, J.S. et al. (2007) Incidence Trends in Primary Malignant Penile Cancer. *Urology Oncology*, vol. 25 (5), pp. 361–367.
10. American Cancer Society (2016) *Can Penile Cancer Be Prevented?* www.cancer.org/cancer/penile-cancer/causes-risks-prevention/prevention.html.
11. Morris, B.J. & Waskett, J.H. (2012) Circumcision Reduces Prostate Cancer Risk. *Cancer*, vol. 14 (5), pp. 661–662.
12. Earp, B. (2015) Do the Benefits of Male Circumcision Outweigh the Risks? A Critique of the Proposed CDC Guidelines. *Frontiers in Pediatrics*, vol. 3 (18), doi:10.3389/fped.2015.00018.
13. Bcheraoui, C. et al. (2014) Rates of Adverse Events Associated with Male Circumcision in US Medical Settings, 2001 to 2010. *JAMA Pediatric*, vol. 168 (7), doi:10.1001/jamapediatrics.2013.5414.
14. The Royal Australasian College of Physicians (2010) Circumcision of Infant

Males. www.racp.edu.au/docs/default-source/advocacy-library/circumcision-of-infant-males.pdf.
15. England, C. (2016) Doctors in Denmark Want to Stop Circumcision for Under-18s. *Independent*, 7 December 2016.
16. Canadian Pediatric Society (2015) Newborn Male Circumcision. Position statement, 8 September 2015, www.cps.ca/en/documents/position/circumcision.
17. As quoted in Wheeler, R. & Malone, P. (2013) Male Circumcision: Risk Versus Benefit. *Archives of Disease in Childhood*, vol. 98 (3), pp. 321–322.
18. Sayre, L.A. (1879) Partial Paralysis from Reflex Irritation, Caused by Congenital Phimosis and Adherent Prepuce. *Transactions of the American Medical Association*, vol. 23 (205).
19. Sayre, L.A. (1879) Partial Paralysis from Reflex Irritation, Caused by Congenital Phimosis and Adherent Prepuce. *Transactions of the American Medical Association*, vol. 23 (205).
20. Kellogg, J.H. (1877) *Plain Facts for Old and Young*. Project Gutenberg.
21. Figures vary around the globe: Shankar, K.R. & Rickwood, A.M. (1999) The Incidence of Phimosis in Boys. *BJU International*, vol. 84 (1), pp. 101–102 conclude: 'The incidence of pathological phimosis in boys was 0.4 cases/1000 boys per year, or 0.6% of boys affected by their 15th birthday, a value lower than previous estimates and exceeded more than eight-fold by the proportion of English boys currently circumcised for "phimosis".'
22. Called the Mathieu procedure, as outlined in Mathieu, P. (1932) Traitement en un temps de l'hypospadias balanique et juxtabalanique. *Journale Chirugerie*, vol. 481.
23. Associated Press (2004) David Reimer, 38, Subject of the John/Joan Case. Via *The New York Times*, 12 May 2004.
24. Mendoza, N. & Motos, N.A. (2013) Androgen Insensitivity Syndrome. *Gynecological Endocrinology*, vol. 29 (1), pp. 1–5, doi:10.3109/09513590.2012.705378.
25. Toerien, M., Wilkinson, S. & Choi, P. (2005) Body Hair Removal: The 'Mundane' Production of Normative Femininity. *Sex Roles*, vol. 52 (5), pp. 399–406, doi:10.1007/s11199-005-2682-5.
26. Rowen, T., Gaither, T. & Mohannad, A. (2016) Pubic Hair Grooming Prevalence and Motivation Among Women in the United States. *JAMA Dermatology*, vol. 152 (10), pp. 1106–1113.
27. Imperato-McGinley, J., Peterson, R.E., Gautier, T. & Sturla, E. (1979) Androgens and the Evolution of Male-gender Identity Among Male Pseudohermaphrodites with 5α-reductase Deficiency. *New England Journal of Medicine*, vol. 300 (22), pp. 1233–1237.
28. Herdt, G. (1990) Mistaken Gender: 5-Alpha Reductase Hermaphroditism and Biological Reductionism in Sexual Identity Reconsidered. *American Anthropologist*, vol. 92 (2), pp. 433–446.
29. Herdt, G. (1994) *Third Sex, Third Gender: Beyond Sexual Dimorphism in Culture and History*. Zone Books.
30. Goldstein, J.M. et al. (2001) Normal Sexual Dimorphism of the Adult Human Brain Assessed by In Vivo Magnetic Resonance Imaging. *Cerebral Cortex*, vol. 11 (6), pp. 490–497.
31. Ritchie, S. et al. (2017) Sex Differences in the Adult Human Brain: Evidence from 5,216 UK Biobank Participants. *BioRxiv*, doi:https://doi.org/10.1101/123729.

32. Tan, A. (2016) The Human Hippocampus Is Not Sexually-dimorphic: Meta-analysis of Structural MRI Volumes. *NeuroImage*, vol. 124, pp. 350–366.
33. Witelson, S.F., Beresh, H. & Kigar, D.L. (2006) Intelligence and Brain Size in 100 Postmortem Brains: Sex, Lateralization and Age Factors. *Brain*, vol. 129 (2), pp. 386–398.
34. Rabinowicz, T. et al. (2002) Structure of the Cerebral Cortex in Men and Women. *Journal of Neuropathology & Experimental Neurology*, vol. 61 (1), pp. 46–57.
35. Nishizawa, S. (1997) Differences Between Males and Females in Rates of Serotonin Synthesis in Human Brain. *PNAS*, vol. 94 (10), pp. 5308–5313.
36. Rametti, G. et al. (2011) White Matter Microstructure in Female to Male Transsexuals before Cross-sex Hormonal Treatment: A Diffusion Tensor Imaging Study. *Journal of Psychiatric Research*, vol. 45 (2), pp. 199–204.
37. Zubiaurre-Elorza, L. et al. (2013) Cortical Thickness in Untreated Transsexuals. *Cerebral Cortex*, vol. 23 (12), pp. 2855–2862, doi:10.1093/cercor/bhs267.
38. Diamond, M. (2013) Transsexuality Among Twins: Identity Concordance, Transition, Rearing, and Orientation. *International Journal of Transgenderism*, vol. 14 (1), pp. 24–38.
39. Hare, L. et al. (2009) Androgen Receptor Repeat Length Polymorphism Associated with Male-to-Female Transsexualism. *Biological Psychiatry*, vol. 65 (1), pp. 93–96, doi:10.1016/j.biopsych.2008.08.033.
40. Robles, R. et al. (2016) Removing Transgender Identity from the Classification of Mental Disorders: A Mexican Field Study for ICD-11. *The Lancet Psychiatry*, vol. 3 (9), pp. 850–859.
41. Love, S. (2016) The WHO Says Being Transgender Is a Mental Illness. But That's About to Change. *The Washington Post*, 28 July 2016.
42. Walpole, H. (1818) *Letters from the Honourable Horace Walpole to George Montagu, Esquire*. Rodwell & Martin.
43. Sayre, L.A. (1869) Three Cases of Lead Palsy from the Use of a Cosmetic Called 'Laird's Bloom of Youth', excerpt from *Transactions of the American Medical Association*. Philadelphia: Collins.
44. Hippocrates' treatise 'Airs, Waters, Places', in Mann, W.N. & Chadwick, J. (1978) *The Medical Works of Hippocrates*. Oxford.
45. From 'Plutarch's Rules for the Preservation of Health: A dialogue'.
46. Galen's 'Hygiene' (*De Sanitate Tuenda*).
47. Glissenti (1609) Disorsi morali. In Duby, G. & Perrot, M. (1992) *A History of Women in the West: Renaissance and Enlightenment Paradoxes*. Harvard University Press.
48. Short, T. (1727) *A Discourse Concerning the Causes and Effects of Corpulency*. London: J. Roberts.
49. Wadd, W. (1816) *Cursory Remarks on Corpulence, or, Obesity Considered as a Disease: With a Critical Examination of Ancient and Modern Opinions, Relative to Its Causes and Cure*. J. Callow, via archive.org.
50. Banting, W. (1864) *Letter on Corpulence, Addressed to the Public*. London: Harrison, via archive.org.
51. Lulu Hunt, P. (1918) *Diet and Health: With Key to the Calories*. Reilly & Lee Co., via archive.org.
52. World Health Organization, based on 2016 figures, last reviewed February 2018, www.who.int/mediacentre/factsheets/fs311/en.

53. Vague, J. (1947) Sexual Differentiation, a Factor Affecting the Forms of Obesity. *Presse Méd.*, vol. 30, pp. 339–340.
54. Fothergill, E. et al. (2016) Persistent Metabolic Adaptation 6 Years after 'The Biggest Loser' Competition. *Obesity*, vol. 24 (8), pp. 1612–1619.
55. Brodesser-Akner, T. (2017) Losing It in the Anti-dieting Age. *New York Times*, 2 August 2017.
56. Ilhan, Z.E. et al. (2017) Distinctive Microbiomes and Metabolites Linked with Weight Loss after Gastric Bypass, but Not Gastric Banding. *The ISME Journal*, doi:10.1038/ismej.2017.71.
57. Pastel, E. et al. (2018) Lysyl Oxidase and Adipose Tissue Dysfunction. *Metabolism*, vol. 78, pp. 118–127.
58. Coutinho, S. et al. (2018) Impact of Weight Loss Achieved Through a Multidisciplinary Intervention on Appetite in Patients with Severe Obesity. *American Journal of Physiology-Endocrinology and Metabolism*, doi:10.1152/ajpendo.00322.2017.
59. Puhl, R. & Heuer, C. (2010) Obesity Stigma: Important Considerations for Public Health. *American Journal of Public Health*, vol. 100 (6), pp. 1019–1028.
60. Roe, J., Fuentes, J.M. & Mullins, M.E. (2012) Underdosing of Common Antibiotics for Obese Patients in the ED. *American Journal of Emergency Medicine*, vol. 30 (7), doi:10.1016/j.ajem.2011.05.027.
61. Andersen, K. et al. (2013) Risk of Arrhythmias in 52 755 Long-distance Cross-country Skiers: A Cohort Study. *European Heart Journal*, vol. 34 (47), pp. 3624–3631.
62. Sussman, S., Lisha, N. & Griffiths, M. (2011) Prevalence of the Addictions: A Problem of the Majority or the Minority? *Evaluation and the Health Professions*, vol. 34, pp. 3–56.
63. Levallius, J., Collin, C. & Birgegård, A. (2017) Now You See It, Now You Don't: Compulsive Exercise in Adolescents with an Eating Disorder. *Journal of Eating Disorders*, vol. 5 (9).
64. Willan, R. (1790) A Remarkable Case of Abstinence. In Hunter, R. & MacAlpine, I. (1963) *Three Hundred Years of Psychiatry, 1535–1860*. Oxford University Press.
65. Hopton, E.N. (2011) Anorexia Nervosa in Adolescent Girls: A Culture-bound Disorder of Western Society? *Social Cosmos*, vol. 2, pp. 176–183.
66. Watters, E. (2010) The Americanisation of Mental Illness. *The New York Times*, 10 January 2010.

## PART V DEAD

1. Numbers 19:11, New American Standard Bible.
2. Puckle, B. (1926) *Funeral Customs: Their Origin and Development*. Library of Alexandria.
3. Morales-Pérez, J. et al. (2017) Funerary Practices or Food Delicatessen? Human Remains with Anthropic Marks from the Western Mediterranean Mesolithic. *Journal of Anthropological Archaeology*, vol. 45, pp. 115–130.
4. From Kinzer, S. (1999) Earthquake in Turkey: The Overview: Turkish Quake Relief Turns to Preventing Epidemics. *The New York Times*, 21 August 1999.

'What is important now is that the bodies be recovered from underneath the ruins so that microbes from the corpses do not spread,' said Senol Ergunsay, a doctor in Gölcük.
5. McComb, K., Baker, L. & Moss, C. (2006) African Elephants Show High Levels of Interest in the Skulls and Ivory of Their Own Species. *Biology Letters*, vol. 2 (1), pp. 26–28.
6. Anderson, J.R. (2010) Pan Thanatology. *Current Biology*, vol. 20 (8), pp. R349–R351.
7. Walker, M. et al. (2012) The Excavation of Buried Articulated Neanderthal Skeletons at Sima de las Palomas. *Quaternary International*, vol. 259, pp. 7–21.
8. For detailed counter-arguments, see Gargett, G. et al. (1989) Grave Shortcomings: The Evidence for Neandertal Burial. *Current Anthropology*, vol. 30 (2), pp. 157–190.
9. Bowler, J., Thorne, A. & Polach, H. (1972) Pleistocene Man in Australia: Age and Significance of the Mungo Skeleton. *Nature*, vol. 240, pp. 48–50.
10. *Epic of Gilgamesh*, tablet VII.
11. *Odyssey*, book XI.
12. MacDougall, D. (1907) Hypothesis Concerning Soul Substance Together with Experimental Evidence of the Existence of Such Substance. *Journal of the American Society for Psychical Research*, vol. 1 (5), pp. 237–244.
13. Gresky, J., Haelm, J. & Clare, L. (2017) Modified Human Crania from Göbekli Tepe Provide Evidence for a New Form of Neolithic Skull Cult. *Science Advances*, vol. 3 (6), doi:10.1126/sciadv.1700564.
14. 9.5 per cent to 12 per cent, according to cpr.heart.org.
15. 6 per cent cardiac arrest survival outside of a hospital environment. Institute of Medicine (2015) *Strategies to Improve Cardiac Arrest Survival: A Time to Act*. Washington, DC: The National Academies Press.
16. Brady, W.J., Gurka, K., Mehring, B., Peberdy, M.A. & O'Connor, R.E. (2011) In-hospital Cardiac Arrest: Impact of Monitoring and Witnessed Event on Patient Survival and Neurologic Status at Hospital Discharge. *Resuscitation*, vol. 82 (7), pp. 845–852.
17. Kirino, T. (2000) Delayed Neuronal Death. *Neuropathology*, vol. 20, pp. S95–97.
18. Zandt, B. et al. (2011) Neural Dynamics during Anoxia and the 'Wave of Death'. *PLOS One*, vol. 6 (7), e22127.
19. Norton, L. et al. (2017) Electroencephalographic Recordings During Withdrawal of Life-sustaining Therapy Until 30 Minutes after Declaration of Death. *Canadian Journal of Neurological Sciences*, vol. 44 (2), pp. 139–145.
20. Academy of Medical Royal Colleges (2008) *A Code of Practice for the Diagnosis and Confirmation of Death*.
21. England, C. (2016) 'Dead' Man Found Alive after Night in Mortuary Fridge. *Independent*, 12 December 2016.
22. Granda, C. (2018) La familia del resucitado: «Tiene hasta las marcas pintadas en el cuerpo para hacerle la autopsia». *La Voz de Asturias*, 8 January 2018.
23. Wakefield, C.E. et al. (2010) Attitudes Toward Organ Donation and Donor Behavior: A Review of the International Literature. *Progress in Transplantation*, vol. 20 (4), pp. 380–391.
24. Kerridge, I.H. et al. (2002) Death, Dying and Donation: Organ Transplantation and the Diagnosis of Death. *Journal of Medical Ethics*, vol. 28 (2), pp. 89–94.

25. Rithalia, A. et al. (2009) Impact of Presumed Consent for Organ Donation on Donation Rates: A Systematic Review. *British Medical Journal*, vol. 338, a3162.
26. Grow, B. & Shiffman, J. (2017) In the U.S. Market for Human Bodies, Almost Anyone Can Dissect and Sell the Dead. *Reuters Investigates*, 24 October 2017.
27. Figures taken from www.aokigaharaforest.com, last accessed 22 February 2018.
28. Hirokawa, S. et al. (2012) Mental Disorders and Suicide in Japan: A Nationwide Psychological Autopsy Case-control Study. *Journal of Affected Disorders*, vol. 140 (2), pp. 168–175.
29. World Health Organization (2017) Suicide Rates, Age-standardized, Data by Country. Global Health Observatory Data Repository.
30. Hackerson, D. (2014) Kamikaze Attacks by the Numbers: A Statistical Analysis of Japan's Wartime Strategy. *The Fair Jilt*, https://thefairjiltarchive.wordpress.com.
31. Economist Films (2015) *24 & Ready to Die*, http://press.economist.com/stories/9535-24-and-ready-to-die-a-documentary-from-economist-films.
32. Henry de Bracton (1235) *De Legibus et Consuetudinibus Angliae* ('On the Laws and Customs of England').
33. *Suicide Act 1961* (9 & 10 Eliz 2 c 60).
34. Ginsberg, M. (1959) *Law and Opinion in England in the 20th Century*. University of California Press.
35. On 17 January 2018, Virginia's legislature voted against repealing the common law of suicide, vote result 2 to 6.
36. Moscicki, E.K. (1997) Identification of Suicide Risk Factors Using Epidemiologic Studies. *The Psychiatric Clinics of North America*, vol. 20 (3), pp. 499–517.
37. Ernst, C. et al. (2004) Suicide and No Axis I Psychopathology. *BMC Psychiatry*, vol. 4 (7).
38. Zhang, J., Xiao, S. & Zhou, L. (2010) Mental Disorders and Suicide among Young Rural Chinese: A Case-control Psychological Autopsy Study. *American Journal of Psychiatry*, vol. 167 (7), pp. 773–781.
39. Hirokawa, S. et al. (2012) Mental Disorders and Suicide in Japan: A Nationwide Psychological Autopsy Case-control Study. *Journal of Affected Disorders*, vol. 140 (2), pp. 168–175.
40. Li, X.Y., Phillips, M.R., Zhang, Y.P., Xu, D. & Yang, G.H. (2008) Risk Factors for Suicide in China's Youth: A Case-control Study. *Psychological Medicine*, vol. 38 (3), pp. 397–406.
41. Manoranjitham, S.D., Rajkumar, A.P., Thangadurai, P., Prasad, J., Jayakaran, R. & Jacob, K.S. (2010) Risk Factors for Suicide in Rural South India. *The British Journal of Psychiatry*, vol. 196 (1), pp. 26–30.
42. Ali, N. et al. (2014) Pattern of Suicides in 2009: Data from the National Suicide Registry Malaysia. *Asia Pacific Psychiatry*, vol. 6, pp. 217–225.
43. Arsenault-Lapierre, G., Kim, C. & Turecki, G. (2004) Psychiatric Diagnoses in 3275 Suicides: A Meta-analysis. *BMC Psychiatry*, vol. 4 (37).
44. Arsenault-Lapierre, G., Kim, C. & Turecki, G. (2004) Psychiatric Diagnoses in 3275 Suicides: A Meta-analysis. *BMC Psychiatry*, vol. 4 (37).
45. Nicolaas Hartsoeker (1694) *Essai de Dioptrique*.
46. Brosens, J. (2014) Uterine Selection of Human Embryos at Implantation. *Scientific Reports*, 3894, doi:10.1038/srep03894.

47. Pliny the Elder, *Natural History*.
48. Koukl, G. (2013) What Exodus 21:22 Says about Abortion. *Stand to Reason*, 4 February 2013, www.str.org/articles/what-exodus-21-22-says-about-abortion#.WZDrz1EjGUk.
49. Clarke, A. (1832) Verse-by-verse Bible Commentary: Psalms 51:5. *Study Light*, www.studylight.org/commentary/psalms/51-5.html.
50. William Blackstone (1770) Blackstone Commentaries. In Dunstan, G.R. & Seller, M.J. (eds) (1988) *The Status of the Human Embryo: Perspectives from Moral Tradition*. King Edward's Hospital Fund for London.
51. Becker, E. (1997) *The Denial of Death*. Free Press.

## PART VI WELL

1. Charman, T. et al. (2011) IQ in Children with Autism Spectrum Disorders: Data from the Special Needs and Autism Project (SNAP). *Psychological Medicine*, 41 (3), pp. 619–627.
2. Treffert, D.A. (2009) The Savant Syndrome: An Extraordinary Condition. A Synopsis: Past, Present, Future. *Philosophical Transactions of the Royal Society B*, vol. 364 (1522), pp. 1351–1357.
3. Rush, B. (1789) Account of a Wonderful Talent for Arithmetical Calculation, in an African Slave Living in Virginia. *American Museum, or Universal Magazine*, January 1789.
4. It's 2,210,500,800, to save you the trouble of reaching for a calculator.
5. Medicine: The Child Is Father. *Time*, 25 July 1960.
6. Stoodley, C. et al. (2017) Altered Cerebellar Connectivity in Autism and Cerebellar-mediated Rescue of Autism-related Behaviors in Mice. *Nature Neuroscience*, vol. 20, pp. 1744–1751.
7. Krishnan, V. et al. (2017) Autism Gene Ube3a and Seizures Impair Sociability by Repressing VTA Cbln1. *Nature*, vol. 543, pp. 507–512.
8. Frietag, C.M. (2007) The Genetics of Autistic Disorders and Its Clinical Relevance: A Review of the Literature. *Molecular Psychiatry*, vol. 12, pp. 2–22.
9. Buja, A. et al. (2017) Damaging De Novo Mutations Diminish Motor Skills in Children on the Autism Spectrum. *PNAS*, doi:https://doi.org/10.1073/pnas.1715427115.
10. Kim, D. et al. (2017) The Joint Effect of Air Pollution Exposure and Copy Number Variation on Risk for Autism. *Autism Research*, vol. 10 (9), pp. 1470–1480; Li, Q. et al. (2017). The Gut Microbiota and Autism Spectrum Disorders. *Frontiers in Cellular Neuroscience*, vol. 11 (120).
11. Koberstein, J.N. et al. (2018) Learning-dependent Chromatin Remodeling Highlights Noncoding Regulatory Regions Linked to Autism. *Science Signalling*, vol. 11 (513), doi:10.1126/scisignal.aan6500.
12. Woodbury-Smith, M.R., Boyd, K. & Szatmari, P. (2010) Autism Spectrum Disorders, Schizophrenia and Diagnostic Confusion. *Journal of Psychiatry & Neuroscience*, vol. 35 (5), p. 360.
13. Goldman, S. (2014) Opinion: Sex, Gender and the Diagnosis of Autism: A Biosocial View of the Male Preponderance. *Research in Autism Spectrum Disorders*, vol. 7 (6), pp. 675–679.

14. Ecker, C. et al. (2017) Association Between the Probability of Autism Spectrum Disorder and Normative Sex-related Phenotypic Diversity in Brain Structure. *JAMA Psychiatry*, vol. 74 (4), pp. 329–338.
15. Chen, F. et al. (2017) Isolation and Whole Genome Sequencing of Fetal Cells from Maternal Blood Towards the Ultimate Non-invasive Prenatal Testing. *Prenatal Diagnosis*, vol. 37, pp. 1311–1321.
16. Galton, F. (1908) *Memories of My Life*. Galton.org.
17. Bell, A. (1884) Lecture titled 'Memoir upon the Formation of a Deaf Variety of the Human Race', National Academy of Sciences, via archive.org.
18. Acts 1907, Laws of the State of Indiana, passed at the Sixty-Fifth Regular Session of the General Assembly, Indianapolis. William B. Burford.
19. *Buck v. Bell*, 274 U.S. 200 (1927).
20. As quoted in a PBS biography, *Freedom Summer*. See www.pbs.org/wgbh/americanexperience/features/freedomsummer-hamer.
21. Committee to End Sterilization Abuse (CESA) (1977) *Sterilization Abuse of Women: The Facts*. Via www.cwluherstory.org/health/sterlization-abuse-a-task-for-the-womens-movement.
22. Ma, H. et al. (2017) Correction of a Pathogenic Gene Mutation in Human Embryos. *Nature*, no. 548, pp. 413–419.
23. From Theophrastus von Hohenheim's *De natura rerum* (1537), a text addressed to his brother late in life. Via archive.org.
24. Gnoth, C. et al. (2003) Time to Pregnancy: Results of the German Prospective Study and Impact on the Management of Infertility. *Human Reproduction*, vol. 18 (9), pp. 1959–1966.
25. Levine, H. et al. (2017) Temporal Trends in Sperm Count: A Systematic Review and Meta-regression Analysis. *Human Reproduction Update*, vol. 23 (6), pp. 646–659.
26. Based on IVF Australia treatment costs, as per www.ivf.com.au/ivf-fees/ivf-costs, last accessed 25 February 2018.
27. Phillips, B., Li, J. & Taylor, M. (2013) Cost of Kids: The Cost of Raising Children in Australia. *AMP.NATSEM Income and Wealth Report*, Issue 33, Sydney, AMP.
28. US Center for Nutrition Policy and Promotion (2017) *Expenditures on Children by Families, 2015*. US Department of Agriculture, Miscellaneous Report No. 1528–2015.
29. England, C. (2017) Indian Woman Who Had Baby at 72 Says She Has No Regrets – but Being a Mother Is Harder than She Expected. *Independent*, 9 March 2017.
30. Allaoui, T. (2016) 63-year-old Woman Gives Birth to IVF Baby in Melbourne. *Herald Sun*, 3 August 2016.
31. AAP (2006) Abbott Backs Away from Restricting IVF. *The Age*, 29 November 2006.
32. As quoted by Alexander, H. (2016) Calls for Restrictions on Medicare Access to IVF Subsidies for Older Women. *The Sydney Morning Herald*, 2 September 2016.
33. Wang, Y.A., Macaldowie, A., Hayward, I., Chambers, G.M. & Sullivan, E.A. (2011) *Assisted Reproductive Technology in Australia and New Zealand 2009*. Assisted Reproduction Technology Series no. 15, cat. no. PER 51. Canberra: AIHW.

34. Tamar, R. & Zil, G. (2018) Case Report: Induced Lactation in a Transgender Woman. *Transgender Health*, vol. 3 (1), pp. 24–26.
35. Brännström, M. et al. (2014) Livebirth after Uterus Transplantation. *The Lancet*, vol. 385 (9968), pp. 607–616.
36. Partridge, E.A. et al. (2017) An Extra-uterine System to Physiologically Support the Extreme Premature Lamb. *Nature Communications*, vol. 8, doi:10.1038/ncomms15112.
37. Liu, S. et al. (2018) Cloning of Macaque Monkeys by Somatic Cell Nuclear Transfer. *Cell*, vol. 172 (4), pp. 881–887.
38. Zhang, J. et al. (2016) First Live Birth Using Human Oocytes Reconstituted by Spindle Nuclear Transfer for Mitochondrial DNA Mutation Causing Leigh Syndrome. *Fertility and Sterility*, vol. 106 (3), pp. e375–376.
39. As quoted in Gibson, O. (2012) Paralympics 2012: Oscar Pistorius Erupts After Alan Oliveira Wins Gold. *The Guardian*, 3 September 2012.
40. Weyand, P.G. et al. (2009) The Fastest Runner on Artificial Legs: Different Limbs, Similar Function? *Journal of Applied Physiology*, vol. 107, pp. 903–911; Potthast, W. & Brüggemann, P. (2010) *Comparison of Sprinting Mechanics of the Double Transtibial Amputee Oscar Pistorius with Able Bodied Athletes*. Institute of Biomechanics and Orthopaedics, German Sport University Cologne, Germany.
41. Ireland, T. (2017) 'I Want to Help Humans Genetically Modify Themselves'. *The Guardian*, 24 December 2017.
42. As quoted by Alba, D. (2016) Zuckerberg and Chan Promise $3 Billion to Cure Every Disease. *Wired.com*, 21 September 2016.
43. IBM (2017) 10 Key Marketing Trends for 2017 and Ideas for Exceeding Customer Expectations. *IBM.com*, www-01.ibm.com/common/ssi/cgi-bin/ssialias?htmlfid=WRL12345USEN.
44. Vlachogiannis, G. (2018) Patient-derived Organoids Model Treatment Response of Metastatic Gastrointestinal Cancers. *Science*, vol. 359 (6378), pp. 920–926.
45. Schork, N.J. (2015) Personalized Medicine: Time for One-person Trials. *Nature*, vol. 520 (7549), doi:10.1038/520609a.
46. Carvalho, T. & Zhu, T. (2014) The Human Genome Project (1990–2003). *Embryo Project Encyclopedia*, http://embryo.asu.edu/handle/10776/7829.
47. Reardon, S. (2015) Obama to Seek $215 Million for Precision-medicine Plan. *Nature News*, 30 January 2015.
48. McCoy, T. et al. (2018) High Throughput Phenotyping for Dimensional Psychopathology in Electronic Health Records. *Biological Psychiatry*, doi.org/10.1016/j.biopsych.2018.01.011.